SpringerBriefs in Applied Sciences and Technology

SpringerBriefs present concise summaries of cutting-edge research and practical applications across a wide spectrum of fields. Featuring compact volumes of 50 to 125 pages, the series covers a range of content from professional to academic.

Typical publications can be:

- A timely report of state-of-the art methods
- An introduction to or a manual for the application of mathematical or computer techniques
- A bridge between new research results, as published in journal articles
- A snapshot of a hot or emerging topic
- An in-depth case study
- A presentation of core concepts that students must understand in order to make independent contributions

SpringerBriefs are characterized by fast, global electronic dissemination, standard publishing contracts, standardized manuscript preparation and formatting guidelines, and expedited production schedules.

On the one hand, **SpringerBriefs in Applied Sciences and Technology** are devoted to the publication of fundamentals and applications within the different classical engineering disciplines as well as in interdisciplinary fields that recently emerged between these areas. On the other hand, as the boundary separating fundamental research and applied technology is more and more dissolving, this series is particularly open to trans-disciplinary topics between fundamental science and engineering.

Indexed by EI-Compendex, SCOPUS and Springerlink.

Mohd Jumain Jalil · Intan Suhada Azmi ·
Norhafini Hambali · Noorfazlida Binti Mohamed ·
Siti Nadia Abdullah · Norin Hafizah Rahim ·
Mohammad 'Aathif Addli

Bio-Based Epoxides

Renewable Materials for Green Chemical Precursors

Mohd Jumain Jalil
Faculty of Chemical Engineering
Universiti Teknologi MARA (UiTM)
Cawangan Johor, Kampus Pasir Gudang
Masai, Johor, Malaysia

Intan Suhada Azmi
Faculty of Chemical Engineering
Universiti Teknologi MARA (UiTM)
Cawangan Johor, Kampus Pasir Gudang
Masai, Johor, Malaysia

Norhafini Hambali
Faculty of Chemical Engineering
Universiti Teknologi MARA (UiTM)
Cawangan Johor, Kampus Pasir Gudang
Masai, Johor, Malaysia

Noorfazlida Binti Mohamed
Faculty of Chemical Engineering
Universiti Teknologi MARA (UiTM)
Cawangan Johor, Kampus Pasir Gudang
Masai, Johor, Malaysia

Siti Nadia Abdullah
Faculty of Chemical Engineering
Universiti Teknologi MARA (UiTM)
Cawangan Johor, Kampus Pasir Gudang
Masai, Johor, Malaysia

Norin Hafizah Rahim
Faculty of Chemical Engineering
Universiti Teknologi MARA (UiTM)
Cawangan Johor, Kampus Pasir Gudang
Masai, Johor, Malaysia

Mohammad 'Aathif Addli
Faculty of Chemical Engineering
Universiti Teknologi MARA (UiTM)
Cawangan Johor, Kampus Pasir Gudang
Masai, Johor, Malaysia

ISSN 2191-530X ISSN 2191-5318 (electronic)
SpringerBriefs in Applied Sciences and Technology
ISBN 978-981-95-5577-2 ISBN 978-981-95-5578-9 (eBook)
https://doi.org/10.1007/978-981-95-5578-9

This Springer imprint is published by the registered company Springer Nature Singapore Pte Ltd.
The registered company address is: 152 Beach Road, #21-01/04 Gateway East, Singapore 189721, Singapore

Preface

This book highlights the emerging potential of bio-based epoxides as sustainable alternatives to petroleum-derived chemical precursors. It explores a range of renewable feedstocks such as vegetable oils, lignocellulosic biomass, and fatty acids, and their conversion into epoxides through eco-friendly processes. Recent advances in catalyst development, green synthesis techniques, and optimization strategies for enhancing oxirane yield and stability are presented, making this work a timely contribution to the field of green chemistry and oleochemical innovation.

In addition, the book emphasizes industrial applications of bio-based epoxides in polymers, coatings, resins, and adhesives while addressing environmental impact, regulatory concerns, and future challenges. By bridging technical insights with sustainability aspects, it serves as a practical reference for researchers, industry professionals, and students who aspire to contribute toward greener chemical manufacturing and the development of a circular bioeconomy.

Chapter 1 related on Palm kernel oil (PKO) is a renewable feedstock widely explored for producing value-added products. Epoxidation is a key modification route, generating oxirane groups useful for polymers, plasticizers, and surfactants while supporting the shift from petroleum-based resources. Despite its potential, challenges such as side reactions and limited stability persist. This chapter introduces the background and significance of PKO epoxidation, setting the stage for discussion on methods, catalysts, and optimization strategies. Chapter 2 examines the use of bio-based epoxides in coatings and adhesives, covering their formulations, performance, and industrial applications. It also considers environmental impact, regulatory requirements, and highlights the challenges and future trends in adopting these sustainable materials.

Chapter 3 discussed kinetic modeling of the epoxidation process, describes the reaction mechanism, rate equations, and kinetic parameters to predict oxirane yield and minimize side reactions. It serves as a tool for optimizing conditions, validating experimental data, and supporting scale-up for industrial applications. Chapter 4 discusses the catalytic epoxidation of waste cooking oil using in situ generated hybrid peracids. It outlines the reaction mechanism, the role of hybrid oxygen carriers, and examines how temperature and hydrogen peroxide molar ratio influence oxirane content.

Chapter 5 discusses extraction of rubber seed oil using a mixed solvent system that provides high oil recovery while maintaining the unsaturation content crucial for epoxidation. This approach ensures sufficient double bonds are preserved, enabling efficient conversion into epoxidized derivatives for polymer and material applications. Chapter 6 examines catalytic epoxidation of palm oil and waste cooking oil using heterogeneous catalysts, offering a greener, reusable, and more selective pathway compared to homogeneous systems. This chapter highlights catalyst types, reaction mechanisms, and performance in improving oxirane yield while reducing side reactions and waste generation. Chapter 7 focuses on the optimization and kinetic modeling of palm stearin epoxidation using an in situ peracid mechanism catalyzed by sulfuric acid. It explains how reaction conditions (temperature, molar ratios, catalyst loading, and time) affect oxirane yield and selectivity, while kinetic modeling is applied to estimate rate constants, describe side reactions such as ring opening, and provide predictive insight for scale-up and process design.

Masai, Malaysia

Mohd Jumain Jalil
Intan Suhada Azmi
Norhafini Hambali
Noorfazlida Binti Mohamed
Siti Nadia Abdullah
Norin Hafizah Rahim
Mohammad 'Aathif Addli

Acknowledgements First and foremost, praise and thanks to God for granting the opportunity, physical and mental strength, and valuable time to complete this book, *Bio-Based Epoxides: Renewable Materials for Green Chemical Precursors*. A special dedication of appreciation is extended to all fellow members for their help, guidance, ideas, and advice throughout the stages of data collection, editing, and preparation of this book.

We would also like to express our sincere gratitude to UiTM Johor, Cawangan Pasir Gudang, for the continuous support and encouragement in making this work possible. Our heartfelt appreciation is extended to colleagues and collaborators whose expertise and assistance have been invaluable over the past years.

Competing Interests The authors have no competing interests to declare that are relevant to the content of this manuscript.

Contents

About the Authors

Ir. Dr. Mohd Jumain Jalil is a lecturer at Universiti Teknologi MARA (UiTM), Cawangan Johor. His research focuses on biomass conversion and green chemical processes. Dr. Jumain has authored several papers in reputable scientific journals. He also supervises undergraduate and postgraduate research projects. e-mail: mjumain0686@uitm.edu.my

Ir. Dr. Intan Suhada Azmi is a senior lecturer in Chemical Engineering at Universiti Teknologi MARA (UiTM), Pasir Gudang Campus, Malaysia. She is a chartered chemical engineer (C.Eng.) recognized by the Institution of Chemical Engineers (IChemE), UK, and a Professional Engineer (Ir.) registered with the Board of Engineers Malaysia (BEM). Her research focuses on the conversion of sustainable vegetable oils into high value-added products such as eco-friendly polyols. Her areas of interest include catalysis, green chemistry, renewable energy, biomass conversion, and optimization of chemical processes. She has published in several high-impact peer-reviewed journals and serves as a reviewer for reputable international journals. e-mail: intansuhada@uitm.edu.my

Norhafini Hambali is a lecturer at the UiTM Pasir Gudang, Johor, specializing in bioprocess engineering, particularly in biopolymer production. Her current research area focuses on catalyst modification to produce biopolymers from epoxide groups. With over five years of experience, she has been involved in natural products extraction, enhancing extraction yield through co-solvent systems and optimization simulations. She has also contributed actively to environmental engineering, particularly in developing natural coagulants and creating natural filtration systems from waste materials. She has several publications in reputable journals highlighting her work in bioprocess engineering, sustainable biopolymer production, and environmental engineering, particularly in fish cage farming areas. e-mail: norhafini@uitm.edu.my

Noorfazlida Binti Mohamed received her Diploma and Bachelor of Chemical Engineering (Hons.) from Universiti Teknologi MARA (UiTM), Shah Alam in 2012. She subsequently completed her Master of Science in Chemical Engineering at

UiTM Shah Alam in 2018 and is currently pursuing her Doctor of Philosophy (Ph.D.) in Chemical Engineering at UiTM Cawangan Pulau Pinang. Her professional experience includes serving as a Research and Development (R&D) Chemist and Researcher at KCC Paints Sdn. Bhd. from 2015 to 2021, followed by a role at Samurai 2K Aerosol Sdn. Bhd. (2022–2024). She is presently the Head of Research and Development at Smart Paint Manufacturing Sdn. Bhd., where she leads innovation in sustainable and high-performance coating technologies.

Siti Nadia Abdullah is a lecturer in Chemical Engineering at Universiti Teknologi MARA (UiTM), Pasir Gudang Campus, Malaysia. Her research focuses on materials and environmental. Her areas of interest include catalysis, green chemistry, renewable energy, biomass conversion, and optimization of chemical processes. She has published in several high-impact peer-reviewed journals and serves as a reviewer for reputable international journals. e-mail: sitinadia@uitm.edu.my

Norin Hafizah Rahim is a lecturer at Universiti Teknologi MARA (UiTM) specializing in process system engineering, particularly in chemical reaction engineering. Her current research area focuses on palm oil, epoxidation and catalytic reaction. She has several publications in reputable journals highlighting her work in each related area of research. e-mail: norinhafizah@uitm.edu.my

Mohammad 'Aathif Addli is a Ph.D. student at Universiti Teknologi MARA. He has already disseminated his research findings through high-impact, peer-reviewed publications in international journals. His current research endeavors are centered on the application of simulation processes, with a specific focus on a novel hybrid kinetic model methodology for the epoxidation process. The significance of this work is underscored by the feature of his paper on the hybrid kinetic model in the *Journal of Polymer Research.*

List of Figures

List of Tables

Chapter 1
Chemical Modification of Palm Kernel Oil via Epoxidation—A Brief Review

1.1 Introduction

The rising environmental challenges, unsustainability, and dependence on fossil-based fuels have driven numerous research efforts toward renewable energy, which is considered as a competitive and cleaner alternative. The growing attention toward clean technologies is due to their bioavailability of raw materials and positive impact on the environment. In economic perspective, it provides more stable pricing and reduce dependence on fossil fuels, while also adding value to agricultural sectors by creating new markets for crops and residues. The types of renewable raw materials vary across different industries, such as plant-based materials (palm kernel oil, coconut oil), agricultural residues (rice husk, bagasse), natural fibers (cotton), marine biomass (algae) [1], and much more. PKO, as a plant oil-based material, has been receiving great attention from researchers as an alternative raw material for different applications. This PKO is obtained from the kernel of oil palm, which has been found to be a major crop and is largely cultivated in Asia, Africa, and Latin America [2, 3]. In Malaysia, from a total of 5.9 million hectares of oil palm plantations, it produces 2.35 million tons of PKO.

The oil palm fruit produced two types of oils which are palm oil (PO) derived from fiber mesocarp and PKO derived from seed of palm fruit as shown in Fig. 1.1. Regarding to Salimon and Salih [4], fresh fruit bunches of oil palm yield approximately 95% crude palm oil (CPO) from the mesocarp and 5% palm kernel oil (PKO) from the kernel. In addition, Babutande et al. [1] stated the palm oil from palm kernel is only 10% of the 79.46 million metric tons of palm oil produced within 2023–2024. The CPO is mainly utilized in food industries, while PKO is used to develop non-edible products for oleochemical manufacturing. The abundant and biodegradable sources led to research and investigation with PKO as its raw material. Previous researches shown the ability of palm kernel oil as raw material in various applications such as cutting lubricant [4, 5], bio-fuel [1, 2, 6, 7], bio-insulator [8, 9], and polyurethane foam [3, 10].

M. J. Jalil et al., *Bio-Based Epoxides*,
SpringerBriefs in Applied Sciences and Technology,
https://doi.org/10.1007/978-981-95-5578-9_1

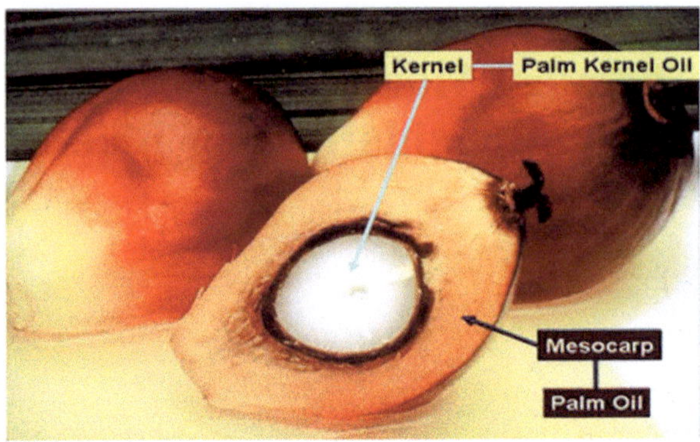

Fig. 1.1 Picture of oil palm

1.2 Chemical Composition of Palm Kernel Oil

PKO is extracted from the kernel of the oil palm fruit (*Elaeis guineensis*), possesses fatty acid composition that differs from many conventional vegetable oils. Its chemical profile is dominantly composed of saturated fatty acids, particularly lauric acid (C12:0), which contributes significantly to its physicochemical characteristics. Tables 1.1 and 1.2 show the physical and chemical properties of PKO. In the context of chemical properties, acid value, free fatty acid (FFA), and peroxide value indicate the quality of oils, in which the lower value indicates better quality of oils and suitability for industrial applications [11]. The acid value and FFA reflect hydrolytic degradation due to moisture, while peroxide value indicates oxidative degradation resulting from exposure to oxygen. According to Kiin-Kabari et al. [11], the maximum value of acid value and FFA acceptable in oils suitable for edibility is 0.4% and 3%, respectively. Based on Table 1.2, PKO has lower values for FFA and acid value, which indicates a huge potential in industrial implementation.

Table 1.3 shows the physical, chemical, and fatty acid properties of PKO, respectively. In general, PKO consists of a complex mixture with a majority of lauric acid (saturated) and oleic acid (unsaturated) [3]. Although PKO contains a relatively low proportion of unsaturated fatty acids, primarily oleic acid (C18:1) and linoleic acid (C18:2), these components are of particular interest in chemical modification

Table 1.1 Physical properties of PKO [11]

Properties	Value
Density (g/ml)	0.884 ± 0.04
Refractive index	1.43 ± 0.02
Slip melting point (°C)	23.67 ± 0.58
Viscosity (pa.S)	0.266 ± 0.03
Flash point (°C)	328.00 ± 9.85
Smoke point	249.99 ± 13.53
Fire point (°C)	291.00 ± 60.70

Table 1.2 Chemical properties of PKO [11]

Properties	Value
Iodine value (g/100 g)	15.13 ± 1.57
Acid value (mgKOH/g)	0.40 ± 0.11
Saponification value (mgKOH/g)	248.33 ± 3.06
Unsaponifiable matter (%)	0.13 ± 0.07
Free fatty acid (%)	0.31 ± 0.04
Peroxide value (meq/kg)	2.01 ± 0.04

Table 1.3 Fatty acid profile of PKO [11]

Fatty acid	Content (%)
Saturated fatty acids	
Lauric (C12:0)	48.00
Myristic (C14:0)	16.20
Stearic (C18:0)	2.50
Palmitic (C16:0)	8.40
Monounsaturated fatty acids	
Oleic (C18:1)	15.00
Linoleic (C18:2)	3.00
Gadoleic (C20:1)	Not dictated
Behenic (C20:0)	0.38

processes such as epoxidation, where the presence of carbon–carbon double bonds is essential. PKO has more composition of lauric (C12:0) and myristic (C14:0) acids with values of 48.00% and 16.20%, respectively, which makes PKO present in solid form at around 20 °C with a low melting point [4, 11]. The degree of saturation in PKO not only influences its thermal and oxidative stability but also determines its reactivity in functionalization reactions. When compared to other oils commonly utilized in epoxidation, such as soybean oil and linseed oil, which are rich in polyunsaturated fatty acids, PKO exhibits a lower unsaturation index. This compositional difference makes PKO less reactive in epoxidation but potentially more stable, highlighting the importance of tailoring processing conditions to the feedstock's unique properties.

1.3 Epoxidation of Palm Kernel Oil

Among the vegetable oils such as soybean oil, canola oil, and linseed oil, PKO is promoted as a leading feedstock with technology enhancement for a wide range of oleochemical applications. The modification of vegetable oils is a crucial process in both industrial and food applications for specific uses due to its characteristics

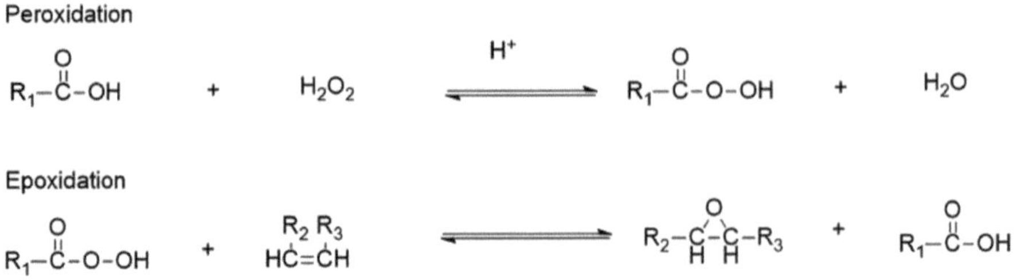

Peroxidation

$$R_1{-}C({=}O){-}OH + H_2O_2 \xrightleftharpoons{H^+} R_1{-}C({=}O){-}O{-}OH + H_2O$$

Epoxidation

$$R_1{-}C({=}O){-}O{-}OH + \overset{R_2\ R_3}{HC{=}CH} \rightleftharpoons R_2{-}\underset{H}{C}{-}O{-}\underset{H}{C}{-}R_3 + R_1{-}C({=}O){-}OH$$

Fig. 1.2 Schematic diagram of peroxidation and epoxidation reaction [12]

of unsatisfactory oxidation stability [12]. Recently, the most extensively applied is chemical modification approach to enhance the reactivity and functionality of vegetable oils. There are various types of chemical modification techniques which include selective hydrogenation, transesterification, epoxidation, and many more. Epoxidation approach involves the conversion of carbon–carbon double bonds in unsaturated fatty acids into oxirane (epoxide) rings. Thus, epoxidation of PKO comprises two levels, which are the conversion to produce epoxide PKO (EPKO) and followed by further conversion to EPKO-based polyol ester [4]. In general, epoxidation mechanism typically involves the reaction of an alkene with an oxygen donor using in situ methods in which peracids are consumed immediately after their formation and form a three-membered epoxide ring [13]. Prileshajev oxidation is the most well-known method in epoxidation, in which the formation of epoxide and carboxylic acid occurs by reaction between vegetable oil and peracid in presence of acid catalyst. The commonly used oxidizing agents are performic acid and peracetic acid, while acid catalysts that can be used in the reaction include acid ion-exchange resin, H_2SO_4, H_3PO_4, HNO_3, and HCl to promote peracid formation [4, 13].

During the epoxidation process, a high content of unsaturated carbon–carbon double bonds is highly desirable. Therefore, the fatty acid composition of vegetable oils can be used to estimate the potential formation of epoxy groups. The presence of epoxy groups in vegetable oils can be monitored directly through oxirane oxygen content analysis and indirectly via iodine value measurements [14]. Epoxidized oils serve as intermediates in polyol production, through direct hydration of oxirane rings. The opening of the oxirane ring can occur through reactions with water, hydrogen chloride or bromide, hydrogenation, or polymerization processes [13]. Figure 1.2 shows the general chemical reaction which comprises peroxidation and epoxidation reaction.

1.4 Comparison of PKO with Other Vegetable Oils

Vegetable oils have emerged as promising renewable feedstocks for the development of bio-based materials, with epoxidation being one of the most widely applied chemical modifications to enhance their reactivity. The efficiency and application potential of epoxidized oils vary significantly depending on the type of oil and its fatty acid composition. There are various types of vegetable oils that are suitable for epoxidation due to the rich content of double bonds, including PKO, soybean oil, canola oil, used cooking oils, mustard oil, and Cardoon seed oil. PKO is extracted from the seed (kernel) of the oil palm fruit that is composed of around 18% oleic acid and widely used in oleochemical industries [4]. While soybean oil is derived from the soybean seed with its high linoleic acid content, extensively applied in the production of epoxidized soybean oil (ESBO) for use in biomaterials industries [15]. The canola oil is a type of vegetable oil that is derived from the seeds of the canola plant, a cultivar of *rapeseed* with high potential for sustainable bio-lubricant development [16]. Used cooking oil can also be used as raw material in epoxidation, produced during household and industrial food preparation.

The used cooking oil mainly consists of triacyl glycerides and can be utilized in the manufacture of biofuels after a suitable pretreatment [17]. For instance, mustard oil is widely used across South Asian countries such as India, popular due to research on its antibacterial, and antioxidant and consists of oleic 15.70% and linoleic 13% [18]. Lastly, cardoon oil used in the epoxidation process because it exhibits characteristics, including a double bond content, that are comparable to those of soybean oil and extracted from the seed of the *Cynara Cardunculus* plant [19]. From the comparison data in Table 1.4, PKO, which is predominantly saturated, contains fewer double bonds and thus exhibits one of the lower epoxidation conversions compared to others. Despite this, its widespread availability in palm-growing regions and potential for use in niche applications such as bio-lubricants and surfactants make it a valuable candidate for targeted chemical modification.

Table 1.4 Comparison data of vegetable oils

Vegetable oils	Dominant unsaturated fatty acids	Epoxide conversion (%)	Reference
Palm kernel oil	Lauric (C12:0), Myristic (C14:0), Oleic (C18:1)	92.5	[4]
Soybean oil	Linoleic (C18:2), Oleic (C18:1), Linolenic (C18:3)	98.6	[15]
Canola oil	Oleic (C18:1), Linoleic (C18:2), Linolenic (C18:3)	95.2	[16]
Used cooking oil	Varies; typically degraded oleic/ linoleic	88.0	[17]
Mustard oil	Erucic (C22:1), Oleic (C18:1), Linoleic (C18:2)	84.0	[18]
Cardoon seed oil	Linoleic (C18:2), Oleic (C18:1)	95.0	[19]

1.5 Application of Palm Kernel Oil in Industries

Recently, numerous researchers have been exploring suitable alternative insulating oils and natural fluids for transformers due to their excellent physicochemical properties [9]. PKO is one of the biodegradable oils that have a huge potential compared to conventional mineral-based insulating fluids in electrical applications. This plant-based liquid insulator and ecofriendly transformers have grown in demand by reducing environmental pollution. The research conducted by Kamhoua et al. [8], studied the combination of palm kernel oil with methyl ester to ensure constant heat transfer and prevention of energy losses in transformers. The combination with methyl ester is to overcome the viscosity problem of vegetable oils through refining and transesterification processes. From the simulation modeling research, this palm kernel oil with methyl ester resulted in maintaining the safe operating temperature at 81.8 °C [8]. In other research, the formation of nanofluid by integrating purified palm kernel oil with aluminum oxide nanoparticles was reported to have good cooling and insulating properties. This is due to its lower viscosity and high breakdown voltage [9].

Besides applying in electrical industries, the utilization of palm kernel oil as bio-lubricant in transportation, machining application, and power generating had attracted significant interest toward researchers [4]. The characteristic green bio-lubricant of epoxidized palm kernel oil-trimethylolpropane polyol ester (EPKO-TMPE) showed good compatibility within the same ISO VG 46 grade as commercial lubricants. This EPKO-TMPE also possesses Newtonian fluid lubricant and is suitable for compressor and turbine applications [4]. Meanwhile, Kazeem et al. [5] analyzed palm kernel oil as cutting lubricant in machining industries using optimization technique. The former lubricant possesses corrosion effect on materials and leads to environmental pollution; hence, the substitution with vegetable-based oil has gained attention. The finding shows that that finishing machining characterization had been improved using PKO as lubricant as compared with mineral oils [5].

Derivation of biofuel from palm kernel oil as raw materials also had gained attention to decrease dependency on fossil fuels usage in transportation industries, consequently, improve the environmental quality. The biodiesel possesses interesting characteristics such as non-toxicity, biodegradable, sulfur and aromatic-free content, thus suitable as an alternative source for diesel engine [2]. Ejeromedoghene [2], had investigated the production of biodiesel using transesterification process with a combination of palm kernel oil and methanol as raw materials. The result showed that the optimum condition was at 15:1 ratio (methanol/oil ratio) for a 98% biodiesel production in the presence of acetic acid catalyst [2]. Meanwhile, the yield of biodiesel resulted in 99.73% by using calcined kernel shell powder with the optimum condition of 1:7.989 v/v of methanol/oil ratio studied by Babutande et al. [1]. In addition, the blending of palm kernel oil with Jatropha curcas oil and methanol also offers good potential for biodiesel and alternative fuel, as investigated by Victoria and Tochukwu [20]. The research outcome shown that the optimal condition that yielded 86.66%

of biodiesel was at molar ratio of 5.81 v/v methanol to oil [20]. Why et al. [7], had developed a combination of palm kernel oil with Jet A-1 fuel. The findings found that at the ratio of 20%:80% of palm kernel oil to Jet A-1 fuel as most promising ration due to its showed good combustion characteristics [7]. Besides using transesterification process, Adzahar et al. [6], had proposed different approaches by using deoxygenation reaction in production of biofuel. The reaction of PKO in the presence of bimetallic nickel–cobalt supported iron-based catalysts had yielded 95% selectivity of biofuel at optimum parameter of 5% wt catalyst loading [6].

The fabrication of bio-based polyols derived from natural oils such as palm kernel oil has gained attention as sustainable feedstock as it offers promising potential due to its high availability and biodegradability. The development of polyurethane from palm kernel oil is used in a variety of applications such as furniture, construction, and transportation industries [3]. The experimental analysis in the formation of rigid polyurethane from palm kernel oil as base materials has been done by Septevani et al. [3]. The research successfully prepared rigid polyurethane foams with optimum condition of combination with 50% petroleum-based polyols. The foams' properties of strength, thermal conductivity, thermal insulation, and stability have been improved [3]. In another research conducted by Trinh et al. [10], analyzed the formation of shape memory polyurethane with introduction of palm kernel oil as bio-based polymer synthesized from renewable resources. The result analysis showed that combination of poly(ethylene glycol) (PEG) and palm kernel oil polyol performed a good shape fixity, about 96% and exhibited excellent shape recovery [10].

1.6 Challenges and Future Perspective on Epoxidation of PKO

Despite the growing interest in the chemical modification of renewable oils, the epoxidation of PKO experiences several challenges that limit its full industrial potential. One of the primary issues is the low level of unsaturation in PKO compared to other vegetable oils like soybean or linseed oil, which reduces the number of reactive sites available for epoxidation and results in a lower epoxide yield. Additionally, traditional epoxidation processes often rely on corrosive acids and hazardous reagents, leading to environmental and safety concerns, including the generation of toxic by-products and difficulty in waste disposal. Looking forward, the integration of green chemistry principles such as solvent-free systems, bio-based catalysts, and process intensification will be critical for improving the sustainability and economic viability of PKO epoxidation. Moreover, future studies should explore advanced functionalization of EPKO for high-performance applications in bioplastics, adhesives, and biomedical materials. These efforts align with global sustainability goals, particularly the United Nations' Sustainable Development Goals (SDGs). The continuity of development on innovation and optimization will ensure that PKO has the potential to become a competitive and sustainable raw material in the bio-based chemical industry.

References

1. E.O. Babatunde et al., Preparation and characterization of palm kernel shell (PKS) based biocatalyst for the transformation of kernel oil to biodiesel. South African J. Chem. Eng. **52**(November 2024), 200–210 (2025). https://doi.org/10.1016/j.sajce.2025.02.008
2. O. Ejeromedoghene, Acid-catalyzed transesterification of palm kernel oil (PKO) to biodiesel. Mater. Today Proc. **47**, 1580–1583 (2021). https://doi.org/10.1016/j.matpr.2021.04.042
3. A.A. Septevani, D.A.C. Evans, C. Chaleat, D.J. Martin, P.K. Annamalai, A systematic study substituting polyether polyol with palm kernel oil based polyester polyol in rigid polyurethane foam. Ind. Crops Prod. **66**, 16–26 (2015). https://doi.org/10.1016/j.indcrop.2014.11.053
4. J. Salimon, N. Salih, Epoxidized Malaysian Elaeis guineensis palm kernel oil trimethylolpropane polyol ester as green renewable biolubricants. Biomass and Bioenergy **175**(June) (2023). https://doi.org/10.1016/j.biombioe.2023.106883
5. R.A. Kazeem et al., Evaluation of palm kernel oil as lubricants in cylindrical turning of AISI 304 austenitic stainless steel using Taguchi-grey relational methodology. Mater. Res. Express **10**(12) (2023). https://doi.org/10.1088/2053-1591/ad11fe
6. N.A. Adzahar, G. AbdulKareem-Alsultan, N.A. Mijan, M.S. Mastuli, H.V. Lee, Y.H. Taufiq-Yap, Effect of catalyst synthesis of bimetallic nickel-cobalt supported iron-based catalysts on converting palm kernel oil into bio-jet fuel via deoxygenation reaction. Energy **314**(November), 2025 (2024). https://doi.org/10.1016/j.energy.2024.133957
7. E.S.K. Why, H.C. Ong, H.V. Lee, W.H. Chen, N. Asikin-Mijan, M. Varman, Conversion of bio-jet fuel from palm kernel oil and its blending effect with jet A-1 fuel. Energy Convers. Manag. **243**(May) (2021). https://doi.org/10.1016/j.enconman.2021.114311
8. A. Kamhoua, G. Mengata Mengounou, L. Monkam, A. Moukengue Imano, Contribution to the use of palm kernel oil methyl esters as liquid bio-insulators in distribution transformers: experimentation and simulation of heat transfer. Results Eng. **22**(November 2023) (2024). https://doi.org/10.1016/j.rineng.2024.102316
9. S.O. Oparanti, A.A. Khaleed, A.A. Abdelmalik, Nanofluid from palm kernel oil for high voltage insulation. Mater. Chem. Phys. **259**(October), 2021 (2020). https://doi.org/10.1016/j.matchemphys.2020.123961
10. N.H. Trinh et al., The effect of polyol types and molecular weight on the shape memory properties of palm kernel oil—based polyurethane. Mater. Today Proc. **17**, 898–904 (2019). https://doi.org/10.1016/j.matpr.2019.06.387
11. D.B. Kiin-Kabari, P.S. Umunna, S.Y. Giami, Physicochemical properties and fatty acid profile of African elemi fruit pulp oil compared with palm kernel oil. Eur. J. Agric. Food Sci. **2**(6), 1–5 (2020). https://doi.org/10.24018/ejfood.2020.2.6.149
12. Y. Zeng et al., A review of chemical modification of vegetable oils and their applications. Lubricants **12**(5) (2024). https://doi.org/10.3390/lubricants12050180
13. G. Lewandowski, M. Musik, K. Malarczyk-Matusiak, Ł. Sałaciński, E. Milchert, Epoxidation of vegetable oils, unsaturated fatty acids and fatty acid esters: a review. Mini. Rev. Org. Chem. **17**(4), 412–422 (2020). https://doi.org/10.2174/1570193X16666190430154319
14. N.S. Sayuti, R. Ali, S. Tuan Anuar, Synthesis and characterization of biobased epoxidized edible oils. Univ. Malaysia Teren. J. Undergrad. Res. **3**(4), 195–206 (2021). https://doi.org/10.46754/umtjur.v3i4.252
15. F. Zhang, Y. Dong, S. Lin, X. Gui, J. Hu, A novel amphiphilic phase transfer catalyst for the green epoxidation of soybean oil with hydrogen peroxide **547**(May) (2023)

16. B. Kamyab, H. Wang, T. Razipour, D.W. Chambers, A.S. Bassi, C. Xu, Production of high-end bio-lubricant products via epoxidation of canola oil trimethylolpropane (COTMP) esters. Ind. Crops Prod. **222**(August) (2024). https://doi.org/10.1016/j.indcrop.2024.119791
17. L.M. Ramírez, J.G. Cadavid, A. Orjuela, M.F. Guti, W.F. Boh, Chemical engineering and processing—process intensification epoxidation of used cooking oils: kinetic modeling and reaction optimization **176**(February) (2022)
18. P.D. Jadhav, A.V. Patwardhan, R.D. Kulkarni, Kinetic study of in situ epoxidation of mustard oil. Mol. Catal. **511**(June) (2021). https://doi.org/10.1016/j.mcat.2021.111748
19. T. Cogliano, V. Russo, K. Eränen, R. Tesser, M. Di Serio, T. Salmi, Epoxidation of vegetable oils in continuous device: kinetics, mass transfer and reactor modelling. Chem. Eng. Sci. **294**(March) (2024). https://doi.org/10.1016/j.ces.2024.120079
20. C. Victoria, K. Tochukwu, Optimized biodiesel production from palm kernel and Jatropha curcas oil blend using KOH-supported calcined animal bone catalyst: a response surface methodology and genetic algorithm-Bayesian hybridization **11**(January) (2025)

Chapter 2
Industrial Application: Bio-Based Epoxides in Coatings and Adhesives

2.1 Sources and Synthesis of Bio-Based Epoxides

2.1.1 Traditional Versus Novel Precursors

Conventional epoxy resin synthesis relies on fossil-based epichlorohydrin and bisphenol A. Bio-based innovations have focused on replacing epichlorohydrin with glycerol-derived epichlorohydrin, which is produced from renewable glycerol, a by-product of biodiesel manufacturing [1]. Additionally, alternative renewable monomers have been introduced, including lignin, tannins, cardanols, rosins, vegetable oils, saccharides, isosorbide, and furan derivatives, which can be epoxidized through chemical or enzymatic methods [1, 2].

2.1.2 Biomass-Derived Examples

Epoxidized soybean oil (ESO) is one of the most commercially available bio-based epoxides, offering a low-cost, non-toxic, and versatile platform for resin formulations, adhesives, and plasticizers [3]. Lignin-derived epoxies combine the aromatic stiffness of lignin with renewable content, producing composites with improved mechanical properties [3]. Tung oil and itaconic acid epoxies have demonstrated thermal and chemical resistance comparable to petroleum-based BPA epoxies, while gallic acid-derived epoxies can achieve higher crosslink densities and superior char yields [2]. More recent developments include sorbitol-derived epoxies, which not only match mechanical performance but also provide inherent flame retardancy [4].

M. J. Jalil et al., *Bio-Based Epoxides*,
SpringerBriefs in Applied Sciences and Technology,
https://doi.org/10.1007/978-981-95-5578-9_2

2.2 Industrial Applications

2.2.1 *Coatings, Adhesive, and Composites*

Bio-based epoxides are widely applied in the coatings, adhesives, and composite sectors due to their low volatile organic compound (VOC) emissions and renewable origins [5, 6]. ESO-based systems have been used as matrix resins, self-healing polymers, and vitrimers, providing tunable mechanical and thermal properties [6]. When combined with hyperbranched curing agents such as tannic acid, ESO-based thermosets can achieve superior tensile strength, adjustable glass transition temperatures, and excellent anti-corrosion resistance, making them suitable for protective coatings and industrial adhesives [6].

2.2.2 *Automative, Electronics, and Construction*

In the automotive industry, bio-based epoxides are used in lightweight composites and high-strength adhesives, contributing to improved fuel efficiency and reduced environmental impact [7]. In electronics, they serve as eco-friendly alternatives for laminates and circuit board coatings, offering good dielectric properties and thermal stability [7]. In construction, bio-based epoxy coatings and adhesives provide sustainable building materials with reduced environmental footprints while maintaining structural integrity [5, 7].

2.2.3 *Market Trends*

Market reports forecast significant growth in bio-based epoxy resin demand due to regulatory pressure, corporate sustainability goals, and consumer preference for eco-friendly materials [5, 8]. Applications in laminates, coatings, adhesives, and composites continue to expand as performance gaps between bio-based and petroleum-based systems narrow. The global bio-based epoxy resins market is projected to surpass USD 270 million by 2031, with major players such as Huntsman leading in market share [8].

2.3 Advantage and Limitations

2.3.1 Advantages

The primary advantage of bio-based epoxides is their renewable origin, which reduces dependency on fossil fuels and lowers greenhouse gas emissions [6, 9]. Many are biodegradable or non-toxic, offering safer handling compared to BPA-based epoxies [3]. Certain epoxidized vegetable oils, such as ESO, are cost-effective and multifunctional, functioning both as a resin component and as a plasticizer [6].

2.3.2 Limitations

Despite their environmental advantages, bio-based epoxies often exhibit lower thermal and mechanical performance compared to petroleum-based counterparts due to the absence of aromatic backbones in many renewable monomers [3, 9]. Additionally, some high-performance bio-based alternatives require complex synthesis steps and costly raw materials, such as purified lignin fractions or itaconic acid [2, 3]. Scaling up production to commercial levels also presents challenges related to raw material availability and processing economics [3].

2.4 Future Prospective

Future research will likely focus on tailoring the chemical structure of bio-based epoxides to match or surpass the performance of petroleum-derived systems. This can be achieved through the incorporation of aromatic monomers, multifunctional curing agents, and nanofillers [4, 6]. Advances in bio-based epichlorohydrin synthesis and the development of fully bio-sourced epoxy-amine thermosets will further enhance sustainability [1, 4]. Improving the scalability and cost-effectiveness of high-performance bio-based epoxies remains essential for broader industrial adoption [3, 4].

2.5 Conclusion

Bio-based epoxides represent a promising class of renewable chemical precursors with applications across multiple industrial sectors, including coatings, adhesives, composites, automotive, electronics, and construction. While challenges remain in achieving performance parity and cost competitiveness with petroleum-based epoxies, continuous innovation in renewable feedstocks, synthesis methods, and formulation strategies will accelerate the transition toward sustainable chemical production.

References

1. X. Wang, H. Li, Y. Zhang, J. Zhu, Progress in sustainable bio-based epoxy technology. Processes **13**(4), 1256 (2023). https://doi.org/10.3390/pr13041256
2. S. Ma, J. Zhu, Biobased epoxies derived from myrcene and plant oil: design and properties. ACS Omega **5**(31), 19719–19728 (2020). https://doi.org/10.1021/acsomega.0c02166
3. S. Ma, X. Liu, Y. Jiang, Z. Tang, C. Zhang, J. Zhu, X. Liu, Fully biobased epoxy resins from fatty acids and lignin. Molecules **25**(5), 1158 (2020). https://doi.org/10.3390/molecules25051158
4. R. Auvergne, S. Caillol, G. David, B. Boutevin, J.P. Pascault, Biobased alternatives to bisphenol A in epoxy–amine thermosets. ACS Sustain. Chem. Eng. **2**(2), 120–135 (2014). https://doi.org/10.1021/sc400289j
5. Transparency Market Research, Bio-based epoxy resin market: overview (2021–2031). Retrieved from https://www.transparencymarketresearch.com/biobased-epoxy-resins-market.html
6. C. Zhang, S. Chen, X. Li, J. Zhu, Sustainable bio-based epoxy resins with tunable thermal and mechanical properties and superior anti-corrosion performance. Polymers **15**(20), 4180 (2023). https://doi.org/10.3390/polym15204180
7. Zicai Chemical, Bio-based epoxy resins: paving the way for a sustainable future in polymer materials (2023). Retrieved from https://www.zicaichemical.com/news/bio-based-epoxy-resins-paving-the-way-for-a-sustainable-future-in-polymer-materials
8. Global Newswire, Global bio-based epoxy resins market forecast (2023). Retrieved from https://www.globenewswire.com/news-release/2023/03/02/2619196/0/en/Global-Bio-Based-Epoxy-Resins-Market-to-Surpass-Valuation-of-US-271-1-Million-by-2031-Huntsman-Set-to-Dominate-Bio-Based-Epoxy-Resins-Market-with-Over-10-Share-Astute-Analytica.html
9. P. Jia, Y. Ma, Q. Kong, L. Xu, Q. Li, Y. Zhou, Progress in development of epoxy resin systems based on biomass resources. Green Materials (2020). https://doi.org/10.1680/jgrma.19.00026

Chapter 3
Kinetic Modeling on Epoxidation Process

3.1 Kinetic Modeling in Epoxidation

Kinetic modeling is a crucial tool in chemical reaction engineering, enabling researchers to quantitatively describe the rate and mechanism of chemical processes [1]. By establishing mathematical relationships between reaction rate, reactant concentrations, temperature, and other process variables, kinetic models serve as a foundation for process design, scale-up, optimization, and control [2]. In catalytic reactions, these models also provide valuable insights into the interaction between reactants and active sites, which is essential for improving catalyst performance and understanding reaction pathways. Without kinetic modeling, process development would rely heavily on empirical trial-and-error approaches, which are time-consuming and economically inefficient.

Epoxidation is a significant chemical transformation, wherein an olefin reacts with an oxidizing agent to form an epoxide—a highly versatile intermediate used in various industrial applications [3]. The reaction is often catalyzed using heterogeneous catalysts such as zeolites, titanium silicates, or modified mesoporous materials to achieve high selectivity and yield. The kinetics of epoxidation reactions can be influenced by several factors, including catalyst acidity, pore structure, oxidant concentration, and reaction medium [4, 5]. Understanding these influences through kinetic modeling allows researchers to identify rate-limiting steps, quantify activation energies, and assess the impact of mass transfer limitations. Such insights are critical for optimizing conditions to enhance epoxide formation while minimizing undesired by-products.

Polyol production from olefin feedstocks generally follows a two-phase pathway: first, the olefin undergoes epoxidation to form the epoxide intermediate, and second, this epoxide is subjected to ring-opening reactions with alcohols, water, or other nucleophiles to yield the polyol [6, 7]. The epoxidation phase determines the yield, purity, and distribution of epoxide species, which directly influence the kinetics and quality of the subsequent polyol formation phase. In other words, the polyol process

M. J. Jalil et al., *Bio-Based Epoxides*,
SpringerBriefs in Applied Sciences and Technology,
https://doi.org/10.1007/978-981-95-5578-9_3

is inherently downstream of epoxidation, relying on the efficiency and selectivity of the first step to ensure desirable product properties such as hydroxyl number, viscosity, and molecular weight distribution.

The production of polyols from epoxides is a multi-step process that may involve competing reactions, side reactions, and complex kinetics. A robust kinetic model for polyol synthesis must account for the influence of temperature, catalyst type, and reactant ratios, as well as possible diffusion effects in heterogeneous systems [8–10]. Since polyol properties are strongly linked to reaction kinetics, precise modeling is essential for tailoring product specifications. In combined epoxidation–polyol processes, the kinetics of both steps need to be well-understood to design an integrated and efficient production route. Any unreacted olefin, by-products, or over-oxidation products formed in the first step may also affect the ring-opening kinetics in the second step.

Various kinetic modeling approaches can be employed for these processes, ranging from simple empirical models like the power-law equation to more mechanistic models derived from proposed reaction mechanisms. For heterogeneous catalytic systems, the Langmuir–Hinshelwood and Eley–Rideal models are often applied to account for adsorption, surface reaction, and desorption phenomena [11]. In cases where mass and heat transfer effects are significant, additional terms may be incorporated to capture external and internal diffusion limitations. The value of kinetic modeling extends beyond the laboratory scale, as accurate models enable process simulation, optimization, and control in industrial settings. By integrating kinetic data into process simulation tools, engineers can predict reactor performance under varying conditions, identify optimal operating points, and evaluate the impact of potential process changes. This predictive capability reduces reliance on costly pilot plant trials and accelerates process development timelines. Moreover, in a sustainability context, kinetic modeling can guide the design of processes that maximize resource efficiency and minimize environmental impact by reducing waste and energy consumption.

In summary, kinetic modeling in epoxidation and polyol processes is not merely an academic exercise but a practical necessity for modern chemical process design. It bridges the gap between experimental data and industrial application, providing a framework to interpret reaction behavior, identify bottlenecks, and optimize performance. The integration of experimental studies with modeling approaches ensures that both fundamental understanding and process efficiency are achieved, paving the way for more sustainable and economically viable chemical manufacturing routes.

3.2 Commonly Used Kinetic Models in Epoxidation Studies

3.2.1 First-Order and Pseudo-First-Order Models

First-order kinetic models describe reactions where the rate is directly proportional to the concentration of a single limiting reactant. In mathematical form, the rate equation is expressed as Eq. 3.1

$$r = -\frac{dC_A}{dt} = kC_A \tag{3.1}$$

where C_A is the reactant concentration and k is the first-order rate constant. This model is often applied in systems where the reaction is elementary and unimolecular with respect to the rate-determining step. In the context of epoxidation, if the reaction rate is governed primarily by the consumption of the olefin or oxidant without significant influence from other reactants or intermediates, a first-order approach may be appropriate.

The pseudo-first-order model is a simplification of more complex reaction systems involving two or more reactants. If one reactant is present in large excess, its concentration remains effectively constant throughout the reaction. This allows the rate law to be reduced to a first-order form with respect to the limiting reactant. For example, in homogeneous epoxidation systems where hydrogen peroxide or tert-butyl hydroperoxide is used in large molar excess, its concentration can be assumed constant, and the rate law becomes dependent only on the olefin concentration.

This simplification offers several practical benefits. It allows for straightforward kinetic data fitting by plotting the natural logarithm of concentration against time, yielding a straight line whose slope corresponds to the apparent first-order rate constant. Moreover, the pseudo-first-order model enables easier determination of activation energy through Arrhenius analysis without the need to vary both reactants' concentrations independently. However, this approach also has limitations. In heterogeneous epoxidation systems, factors such as mass transfer, adsorption equilibria, and catalyst deactivation may cause deviations from ideal first-order behavior. Furthermore, when the excess reactant begins to deplete or participates in side reactions, the pseudo-first-order assumption may no longer hold. Therefore, careful experimental design and validation are required before adopting this simplification.

In polyol production, pseudo-first-order kinetics can sometimes describe the ring-opening step if the nucleophile (such as water or alcohol) is in large excess relative to the epoxide. This is particularly useful for designing integrated epoxidation–polyol processes, where the excess nucleophile ensures complete conversion of epoxides while simplifying kinetic modeling.

3.2.2 Langmuir–Hinshelwood (L–H) Model

The Langmuir–Hinshelwood model describes heterogeneous catalytic reactions where both reactants are adsorbed onto the catalyst surface before undergoing a surface reaction. The process typically involves adsorption of species A and B onto active sites, reaction between the adsorbed species, and desorption of the product. The derived rate expression reflects the balance between adsorption and reaction, with the numerator representing the product of adsorption and surface reaction, while the denominator accounts for site coverage by all species. This model is generally applied when both reactants have significant interactions with the catalyst surface, making it suitable for processes such as oxidation, hydrogenation, and epoxidation. This leads to a rate equation of the form:

$$r = \frac{kK_A C_A K_B C_B}{(1 + K_A C_A + K_B C_B)^2}$$

where K_A and K_B are adsorption equilibrium constants for reactants A and B, respectively.

In epoxidation catalyzed by materials such as Ti–SiO_2 or modified ZSM-5, the L–H model provides a mechanistic interpretation of how reactants interact with active sites. For example, in olefin epoxidation using hydrogen peroxide, both the olefin and oxidant may adsorb onto distinct active sites, and the reaction proceeds via a surface peroxide species reacting with the olefin. The L–H framework captures this dual adsorption phenomenon, allowing researchers to determine adsorption constants and identify competitive adsorption effects.

One advantage of the L–H model is that it explicitly accounts for adsorption competition, which can be significant in systems where by-products or solvents also bind to the active sites. This makes the model valuable for interpreting deviations from simple power-law behavior, especially at high reactant concentrations where site saturation occurs. Additionally, L–H kinetics can help distinguish whether surface reaction or adsorption/desorption steps are rate-limiting.

However, applying the L–H model requires accurate adsorption data, which can be challenging to obtain experimentally. Adsorption equilibria may be influenced by temperature, catalyst morphology, and surface heterogeneity, complicating parameter estimation. Furthermore, if one reactant does not adsorb strongly, the L–H mechanism may overcomplicate the system, making the Eley–Rideal model more appropriate. For polyol synthesis, the L–H model is relevant when solid acid catalysts are used for epoxide ring-opening. In such cases, both the epoxide and nucleophile may adsorb on the catalyst surface before undergoing nucleophilic attack, making the model suitable for integrated kinetic descriptions of the two-step process.

3.2.3 Eley–Rideal (E–R) Model

The Eley–Rideal (E–R) model offers an alternative to the L–H mechanism. The Eley–Rideal model, on the other hand, assumes that only one reactant is adsorbed onto the catalyst surface, while the second reactant remains in the bulk phase and reacts directly with the adsorbed species. The reaction mechanism therefore involves adsorption of species A, followed by its direct interaction with species B from the fluid phase. The resulting rate expression reflects adsorption of one species and a direct reaction step, without requiring adsorption of both reactants. This model is more appropriate in cases where only one reactant shows strong adsorption, while the other is weakly bound or remains primarily in the gas or liquid phase. The general rate expression can be written as Eq. 3.2

$$r = \frac{kK_A C_A C_B}{1 + K_A C_A} \tag{3.2}$$

where C_A represents the concentration of the adsorbed reactant, and C_B is the bulk-phase reactant.

In epoxidation systems, the E–R model may apply when the oxidant strongly adsorbs to the catalyst surface, forming an active oxygen species, while the olefin reacts directly from the liquid phase without significant adsorption. This scenario is observed in some titanium-based catalysts, where hydrogen peroxide binds to surface Ti sites, and the olefin approaches from the bulk to undergo oxygen transfer.

The E–R mechanism simplifies the description of systems where adsorption of one reactant is negligible or energetically unfavorable. It is particularly advantageous when experimental adsorption isotherms show strong binding for one reactant and negligible uptake for the other. This reduces the number of kinetic parameters to be estimated and can lead to more robust model fitting. Nevertheless, the E–R model has limitations. It may fail to capture complex adsorption–desorption equilibria or account for situations where both reactants weakly adsorb. Moreover, in processes with competitive adsorption between reactants and solvents, E–R assumptions might oversimplify the system.

For polyol synthesis, an E–R-type mechanism could be relevant in heterogeneous ring-opening reactions if, for example, the epoxide adsorbs on an acidic catalyst site while the nucleophile attacks directly from the solution phase. This mechanistic insight can help design catalysts that selectively promote the desired reaction pathway.

3.2.4 Power Law/Empirical Models

Power-law or empirical kinetic models are widely used when the mechanistic pathway of a reaction is not fully understood, or when a simple, flexible model is needed for process optimization. The general form of the power-law equation in Eq. 3.3.

$$r = kC_A^m C_B^n \tag{3.3}$$

where m and n are empirical reaction orders determined experimentally.

In epoxidation studies, power-law models can be particularly useful during the early stages of kinetic investigation, allowing researchers to quickly quantify the influence of reactant concentrations and temperature on the reaction rate. These models require minimal assumptions about adsorption or surface reaction mechanisms, making them faster to apply than L–H or E–R approaches. Although empirical models lack direct mechanistic interpretation, they are valuable for reactor design and process control, especially when rapid decision-making is needed. By fitting rate data to a power-law form, process engineers can predict performance under different operating conditions and identify optimal feed ratios for maximum conversion or selectivity. However, power-law models have limited predictive power outside the experimental range of concentrations and temperatures. They may also mask underlying phenomena such as catalyst deactivation, mass transfer limitations, or competitive adsorption. Consequently, while power-law models are useful for screening and optimization, they are often replaced by mechanistic models once more information is available.

In polyol synthesis, empirical models can describe the rate of epoxide ring-opening in complex mixtures where multiple alcohols or polyfunctional nucleophiles are present, and the detailed mechanism is challenging to resolve. This approach enables rapid process adjustments while mechanistic studies are still ongoing.

3.2.5 Consecutive Kinetic Model

In epoxidation and ring-opening studies, many systems are better described by a consecutive kinetic model rather than a simple single-step reaction. This approach recognizes that an intermediate is formed before the final product, following the sequence:

$$[\text{Epoxide}]_A \rightarrow^{k_1} [\text{Epoxide} - \text{Polyol}]_B \rightarrow^{k_2} [\text{Polyol}]_C$$

where A corresponds to the epoxy functional groups, B represents the intermediate, and C denotes the final polyol product. The experimental observable, $X(t)$, corresponds to the concentration of the reacting species as a function of time. The kinetic model is mathematically expressed as:

$$X(t) = A\left(e^{ek_1 t} - e^{-k_2 t}\right)$$

where $X(t)$ is the concentration of the intermediate species, A is a proportionality constant reflecting the maximum achievable concentration difference, and k_1 and $k_2(\text{min}^{-1})$ are the rate constants of the first and second steps, respectively. Experimental data were collected as concentration of epoxy groups (mmol/L) versus reaction time (min). The base equations are:

$$\frac{dA}{dt} = -k_1 A, \quad \frac{dB}{dt} = k_1 A - k_2 B, \quad \frac{dC}{dt} = k_2 B$$

With initial conditions of all species equal to 0.

Thus, solving these gives:

$$A(t) = A_0 e^{-k_1 t}$$

$$B(t) = \frac{A_0}{k_2 - k_1}\left(e^{-k_1 t} - e^{-k_2 t}\right)$$

$$C(t) = A_0\left[1 - \frac{k_2 e^{-k_1 t} - k_1 e^{-k_2 t}}{k_2 - k_1}\right]$$

These formulas therefore capture both the mechanistic details and the kinetic behavior, providing a solid basis for model fitting and interpretation.

3.3 Discussion on R^2 Value as a Goodness-Of-Fit Indicator

The coefficient of determination (R^2) is a common statistical measure used in kinetic modeling to quantify the proportion of variance in experimental data that is explained by the model, with values approaching 1 indicating strong agreement between predicted and observed results. Its main advantage lies in its simplicity and intuitive interpretation, making it a quick indicator of model performance. However, R^2 alone cannot confirm the correctness of a kinetic model, as high values may result from overfitting or coincidental correlations that lack mechanistic validity. Furthermore, R^2 does not penalize for the inclusion of unnecessary parameters and can be misleading in nonlinear systems or when the data range is narrow. To achieve more robust model evaluation, R^2 should be complemented with additional statistical and diagnostic tools such as adjusted R^2 (which accounts for the number of model parameters), root mean square error (RMSE) for absolute prediction accuracy, Akaike Information Criterion (AIC) for model parsimony, and residual analysis to detect systematic deviations. Using these metrics in combination ensures a more balanced assessment, capturing both the statistical fit and the mechanistic plausibility of kinetic models.

In practice, robust kinetic model validation in epoxidation and polyol processes should integrate these statistical tools rather than relying on a single indicator. R^2

and adjusted R^2 provide a quick initial measure of fit quality, while RMSE quantifies the average prediction error in physical units that are directly interpretable for process design. AIC assists in selecting the most parsimonious model that balances accuracy and complexity, reducing the risk of overfitting. Residual analysis complements these metrics by revealing systematic deviations that numerical indicators may overlook, such as curvature, heteroscedasticity, or time-dependent trends. By evaluating a kinetic model through this multi-criteria approach, researchers can ensure that it not only matches experimental data statistically but also maintains mechanistic plausibility and predictive reliability across varying process conditions as shown in Table 3.1.

Among the kinetic models applied to epoxidation and ring-opening reactions, the consecutive first-order model consistently provides the best statistical agreement with experimental data. This model separates the slow nucleophilic initiation step (k_1) from the faster propagation step (k_2), giving a more realistic description of the reaction sequence compared to single-step models. In the present study, fitting ECO ring-opening data with the consecutive model yielded very high coefficients of determination ($R^2 \approx 0.99$) and residuals narrowly distributed within ± 0.03, with no systematic trends. This indicates that the model captures the mechanistic behavior accurately while ensuring statistical robustness. In contrast, simpler first-order fits often underestimate early-stage deviations or overpredict final conversion. Thus, the

Table 3.1 Differences of metrics

Metric	Purpose	Advantages	Limitations
R^2 (Coefficient of determination)	Measures proportion of variance in data explained by the model	Simple to calculate; intuitive interpretation	Does not confirm mechanistic validity; can be inflated by overfitting; insensitive to bias
Adjusted R^2	Adjusts R^2 for the number of predictors or parameters	Penalizes unnecessary model complexity; better for comparing models	Can still be misleading if assumptions are violated
RMSE (Root mean square error)	Measures average magnitude of prediction error	Reflects absolute accuracy; same units as data	Sensitive to outliers; does not indicate bias direction
AIC (Akaike information criterion)	Compares model quality while penalizing complexity	Balances goodness of fit and simplicity; useful for model selection	Relative measure; absolute values have no direct meaning
Residual analysis	Examines difference between observed and predicted values	Detects systematic deviations, bias, or poor model assumptions	Requires visual inspection; subjective interpretation possible

goodness of fit indicators confirm that the consecutive model is the most appropriate framework for describing the kinetics of ECO epoxidation and ring-opening, balancing both mechanistic accuracy and predictive reliability.

3.4 Case Examples

3.4.1 Kinetic Modeling in Epoxidation of Palm Oleic Acid

In the study by Azmi [12], palm oleic acid was used as the substrate to produce epoxidized palm oleic acid (EPOA) via an in situ performic acid (HCOOOH) epoxidation mechanism. The research first focused on identifying the optimal hydrogen peroxide concentration to maximize the relative conversion to oxirane (RCO), with experiments conducted at 30%, 35%, and 50% H_2O_2 under fixed conditions of 60 °C, 300 rpm stirring, and a 35 min reaction time. A comprehensive kinetic model was then developed to describe both the epoxidation and oxirane degradation processes, building upon kinetic frameworks for palm kernel oil and employing the differential method with numerical integration via the Runge–Kutta approach. The kinetic parameters were estimated using genetic algorithms implemented in MATLAB, enabling efficient optimization by simultaneously searching for the ideal process variable values and fitting them to the experimental data. Comparison of experimental oxirane content profiles with simulated results showed excellent agreement across all peroxide concentrations, yielding estimated R^2 values of approximately 0.97 for 30% H_2O_2, 0.99 for 35% H_2O_2, and 0.94 for 50% H_2O_2.

The highest fit quality at the intermediate peroxide concentration reflects the model's ability to accurately capture the balance between epoxidation and degradation rates under optimal oxidant conditions. The slight reduction in R^2 at 50% H_2O_2 likely arises from the increased influence of side reactions such as oxirane ring-opening, peroxide decomposition, and possible non-ideal mass transfer behavior at elevated oxidant levels. Overall, the study demonstrated that the proposed kinetic model, supported by advanced parameter estimation techniques, not only provides a statistically robust description of the epoxidation process but also delivers kinetic parameters suitable for scaling up epoxide production to industrial levels.

3.4.2 Kinetic Modeling in Epoxidation of Melon Seed Oil

In another study by Nwosu [13], response surface methodology (RSM) and analysis of variance (ANOVA) were used to optimize the epoxidation of melon seed oil methyl esters. The second-order polynomial model developed achieved very high statistical indicators, with $R^2 = 0.9996$, adjusted $R^2 = 0.9992$, and predicted $R^2 = 0.9944$, showing excellent agreement between experimental and predicted oxirane values.

The model predicted an optimal oxirane content of 3.963% at a stirring speed of 597.821 rpm, reaction time of 3.96 h, and temperature of 71.71 °C, with a desirability index of 1.000. An adaptive neuro-fuzzy inference system (ANFIS) model was also tested, achieving R^2 values of 0.9996 for RSM and 0.9653 for ANFIS, with a mean-squared error as low as 1.99×10^{-6}. While these R^2 values show that the models fit the data extremely well, such near-perfect scores should be interpreted with care.

In some cases, high R^2 can result from overfitting—when the model is tuned too closely to the data and may not perform as well under new conditions. The lower R^2 for ANFIS compared to RSM suggests that the choice of model type and settings can influence predictive accuracy. Although the predicted R^2 for RSM (0.9944) suggests good generalization within the experimental range, further testing outside this range would be important to confirm the model's reliability for scale-up or industrial application.

3.4.3 *Kinetic Modeling in Epoxidation of Grape Seed Oil*

In a study by Catalá et al. [14], the kinetics of grape seed oil epoxidation in supercritical CO_2 were modeled using a pseudo-first-order approach, with iodine value measurements serving as an indicator of double bond concentration. Linear regression of experimental data at 40, 50, and 60 °C yielded coefficients of determination (R^2) of 0.9851, 0.9401, and 0.9818, respectively, demonstrating excellent agreement with the kinetic model. The authors noted that the slightly lower R^2 at 50 °C could be attributed to side reactions or minor experimental variability. The corresponding rate constants increased with temperature, confirming the strong temperature dependence of the reaction. These findings indicate that, at least in the early stages, epoxidation under supercritical conditions can be effectively described using a pseudo-first-order kinetic model, providing a useful framework for process analysis and optimization.

3.4.4 *Kinetic Modeling in Epoxidation of Castor Oil*

In a kinetic modeling study by Addli et al. [15] on castor oil epoxidation at 65 °C, three computational optimization techniques—Particle Swarm Optimization (PSO), Simulated Annealing (SA), and a hybrid PSO–SA algorithm—were applied to predict the temporal evolution of oxirane oxygen content. The modeling employed a pseudo-first-order kinetic framework, justified by the excess concentration of peracetic acid relative to the unsaturated sites in the castor oil, which allowed the reaction rate to be expressed as dependent only on the oxirane oxygen concentration. The model encompassed the key reaction steps of peracetic acid synthesis and decomposition, epoxidation, and in situ ring-opening by water.

The accuracy of each optimization approach was evaluated by comparing simulated results to experimental data, with the coefficient of determination (R^2) serving as the primary metric. The hybrid PSO–SA model achieved the highest accuracy, with an R^2 of 0.9961, indicating excellent agreement with experimental measurements and strong capability in representing both the formation and decline of oxirane oxygen content. The standalone PSO model followed with an R^2 of 0.9836, still demonstrating good predictive performance but showing greater deviation at later stages of the reaction. The SA model recorded the lowest R^2 of 0.9779, reflecting limitations in capturing the full reaction profile.

The enhanced predictive precision of the hybrid PSO–SA model can be attributed to its ability to combine PSO's global exploration capacity with SA's local refinement capability. While PSO rapidly identifies promising regions in the parameter space, SA enables escape from local minima, preventing premature convergence and allowing for more accurate parameter estimation within the pseudo-first-order kinetic structure. These results highlight the effectiveness of hybrid optimization methods in refining kinetic models for complex, multiphase epoxidation systems, where accurate prediction of both formation and degradation steps is crucial for process optimization.

3.4.5 Kinetic Modeling in Epoxidation of Oleic Acid

In the study by Sawitri et al. [1], the kinetics of in situ epoxidation of oleic acid were investigated using a pseudo-homogeneous model, which assumes that oleic acid, aqueous hydrogen peroxide, and acetic acid form a single, well-mixed liquid phase. Four kinetic model variations were developed to describe the system: Model 1 considered the irreversible ring-opening of the oxirane ring by water and formic acid ($R^2 = 0.907$); Model 2 was similar to Model 1 but accounted for the ring-opening reaction as reversible ($R^2 = 0.988$); Model 3 irreversible and reversible ring-opening reactions occurring alongside epoxidation ($R^2 = 0.939$); and Model 4 represented the most comprehensive pathway, incorporating peracetic acid formation, epoxidation, and both irreversible and reversible ring-opening by water and formic acid ($R^2 = 0.990$). The results show a clear improvement in predictive accuracy as additional mechanistic steps were incorporated, with Model 4 providing the best agreement with experimental data. These findings confirm that modeling reversible ring-opening within the pseudo-homogeneous framework yields a more complete representation of oleic acid epoxidation kinetics.

3.4.6 Kinetic Modeling in Epoxidation of Palm Stearin

In the study by Rahim et al. [16], the kinetics of catalytic epoxidation of palm stearin were modeled using a pseudo-homogeneous framework and parameter estimation via simulated annealing combined with the Runge–Kutta numerical integration method.

The model described the formation and decomposition of performic acid, epoxidation of palm stearin, and degradation of the oxirane ring. The simulated epoxide concentration profile at 100 °C showed a reasonable match with experimental results, although deviations were observed between 30 and 40 min due to the idealized assumptions of the model, such as neglecting heat loss, mass transfer limitations, and side reactions. The model achieved a coefficient of determination ($R^2 = 0.52$), reflecting moderate predictive capability and indicating that further refinement such as incorporating non-idealities and additional reaction pathways would be necessary to improve accuracy as shown in Table 3.2.

3.5 Conclusion and Perspectives

Kinetic modeling of epoxidation processes provides critical insights for catalyst design, process optimization, and scale-up. While simple models offer ease of application, complex models can capture mechanistic detail in heterogeneous catalysis. The R^2 value remains a convenient indicator of fit quality but should not be used in isolation. Future research should integrate kinetic modeling with in situ spectroscopy, computational chemistry, and machine learning approaches.

Table 3.2 Case examples comparisons

Feedstock/ system	Kinetic model applied	Key features/ assumptions	Optimization/ method used	R^2 (Goodness of fit)	Remarks
Grape seed oil	Pseudo-first-order	Iodine value as double bond indicator; supercritical CO_2 medium	Linear regression at 40–60 °C	0.9851 (40 °C); 0.9401 (50 °C); 0.9818 (60 °C)	Strong temperature dependence; slightly lower accuracy at 50 °C due to side reactions or variability
Castor oil	Pseudo-first-order (peracetic acid in excess)	Accounts for peracetic acid formation/ decomposition, epoxidation, in situ ring-opening	Particle Swarm Optimization (PSO), Simulated Annealing (SA), Hybrid PSO–SA	PSO–SA = 0.9961; PSO = 0.9836; SA = 0.9779	Hybrid PSO–SA gave best performance; combines global + local search to avoid premature convergence
Oleic acid	Pseudo-homogeneous (4 model variations)	Assumes single liquid phase; included epoxidation + reversible/ irreversible ring-opening	Comparative mechanistic modeling	Model 1 = 0.907; Model 2 = 0.988; Model 3 = 0.939; Model 4 = 0.990	Accuracy improved as more mechanistic steps were included; Model 4 best fit with reversible ring-opening
Palm stearin	Pseudo-homogeneous	Performic acid formation/ decomposition, epoxidation, oxirane degradation	Simulated Annealing + Runge–Kutta	0.52	Only moderate predictive capability; further refinement needed

References

1. D.R. Sawitri, P. Mulyono, Rochmadi, A. Hisyam, A. Budiman, Kinetic investigation for in-situ epoxidation of unsaturated fatty acid based on the pseudo-steady-state-hypothesis (PSSH). J. Oleo Sci. **69**(10), 1297–1305 (2020). https://doi.org/10.5650/jos.ess20034
2. M.A. Addli, I.S. Azmi, M.J. Jalil, In Situ epoxidation of castor oil via synergistic sulfate-impregnated ZSM-5 as catalyst. J. Polym. Environ. **32**(4), 1593–1601 (2024). https://doi.org/10.1007/s10924-023-03056-w
3. T. Saurabh, M. Patnaik, S.L. Bhagt, V.C. Renge, *Epoxidation of Vegetable Oil; A Review*, vol. II (2011)

4. V. Smeets, E.M. Gaigneaux, D.P. Debecker, *Titanosilicate Epoxidation Catalysts: A Review of Challenges and Opportunities* (John Wiley and Sons Inc., 2022) https://doi.org/10.1002/cctc.202101132
5. M. Bocus, S.E. Neale, P. Cnudde, V. Van Speybroeck, Dynamic evolution of catalytic active sites within zeolite catalysis, in *Comprehensive Inorganic Chemistry III* (Elsevier, 2023), pp. 165–200. https://doi.org/10.1016/B978-0-12-823144-9.00012-1
6. M.A. Mohd Noor et al., Molecular weight determination of palm olein polyols by gel permeation chromatography using polyether polyols calibration. J. Am. Oil Chem. Soc. **93**(5), 721–730 (2016). https://doi.org/10.1007/s11746-016-2812-y
7. M.Z. Arniza et al., Synthesis of transesterified palm olein-based polyol and rigid polyurethanes from this polyol. J. Am. Oil Chem. Soc. **92**(2), 243–255 (2015). https://doi.org/10.1007/s11746-015-2592-9
8. E.N. Jacobsen, Asymmetric catalysis of epoxide ring-opening reactions. Acc. Chem. Res. **33**(6), 421–431 (2000). https://doi.org/10.1021/AR960061V/ASSET/IMAGES/MEDIUM/AR960061VU00003A.GIF
9. A. Lidskog, Y. Li, K. Wärnmark, Asymmetric ring-opening of epoxides catalyzed by Metal–Salen Complexes. Catalysts **10**(6) (2020). https://doi.org/10.3390/catal10060705
10. A. Tardy, J. Nicolas, D. Gigmes, C. Lefay, Y. Guillaneuf, Radical ring-opening polymerization: scope, limitations, and application to (bio)degradable materials. Chem. Rev. **117**(3), 1319–1406 (2017). https://doi.org/10.1021/ACS.CHEMREV.6B00319
11. R.J. Baxter, P. Hu, Insight into why the Langmuir-Hinshelwood mechanism is generally preferred. J. Chem. Phys. **116**(11), 4379–4381 (2002). https://doi.org/10.1063/1.1458938
12. I.S. Azmi, M.H.A. Bakar, D.N.A. Raofuddin, H.H. Habri, M.H.M. Azmi, M.J. Jalil, Synthesis and kinetic model of oleic acid-based epoxides by in situ peracid mechanism. Kemija u industriji (3–4, Mar) (2022). https://doi.org/10.15255/KUI.2021.024
13. K. Nwosu-Obieogu, C. Goodnews, G.W. Dzarma, C. Ugwuodo, O. Gabriel, Azadirachta indica seed oil epoxidation process using carbonized melon seed peel catalyst; genetic algorithm coupled artificial neural network approach. S. Afr. J. Chem. Eng. **49**, 258–272 (2024). https://doi.org/10.1016/j.sajce.2024.06.005
14. J. Catalá, J.M. García-Vargas, M.J. Ramos, J.F. Rodríguez, M.T. García, Kinetics of grape seed oil epoxidation in supercritical CO_2. Catalysts **11**(12), 1490 (2021). https://doi.org/10.3390/catal11121490
15. M. 'Aathif Addli, I.S. Azmi, S.D. Nurherdiana, M.A. Ahmad, M.J. Jalil, Ring opening of epoxidized castor oil with applied hybrid kinetic modelling model of particle swarm & simulated annealing. J. Polym. Res. **32**(4), 127 (2025). https://doi.org/10.1007/s10965-025-04352-w
16. N.H. Rahim, M.J. Jalil, N.M. Mubarak, I.S. Azmi, G. Anbuchezhiyan, Catalytic epoxidation of unsaturated fatty acids in palm stearin via in situ peracetic acids mechanism. Sci. Rep. **15**(1), 4789 (2025). https://doi.org/10.1038/s41598-025-89399-x

Chapter 4
Epoxidation of Waste Cooking Oil Applied Hybrid Peracids Mechanism with Applied Titanium Dioxide

4.1 Introduction

Waste cooking oil is the end-product of frying foods using cooking oil which contains vegetable or animal fats that have been processed [1]. Cooking oil is a glycerol ester consisting of different types of essential fatty acids depending on the vegetable oil involved. Waste is categorized as fat and grease and in liquid shape at room temperature [2]. This type of waste is produced by food premises or restaurants, food industry, and household occurring from the preparation of food [3]. Nevertheless, due to its characteristic of not being able or not being soluble in water, it becomes a contaminant to the environment. To reduce the environmental and health impacts of waste cooking oil, it is important to properly dispose of it and one of the alternative methods is epoxidation of waste cooking oil [4]. Epoxidation of palm waste cooking oil is a process that involves converting unsaturated fatty acids in the waste cooking oil into epoxy fatty acids. This reaction is typically performed using an oxidizing agent, such as a peracid, which converts the unsaturated fatty acids into epoxides [5]. The use of palm waste cooking oil for epoxidation has several benefits. Firstly, it allows for the utilization of a waste product, reducing the amount of waste generated by the food industry. Secondly, epoxidation can improve the functional properties of the oil, making it more suitable for use in various applications, such as the production of biodegradable plastics and other materials [6]. Thus, it is necessary to optimize the epoxidation process to improve the efficiency and selectivity of the reaction [7].

The collection and transport systems for collecting waste cooking oil face a challenge, specifically in terms of managing this type of waste. This is due to waste cooking oil coming in the liquid or aqueous form [8]. If carelessly managed, consequently it will end up in the sewage system, water and will further damage the environment. Waste cooking oil is harmful or toxic to the environment and health of the consumers. The use of waste from cooking oil is believed will lead to cause cancer due to the toxic contents produced when the oil is oxidized from the fried foods [9]. Most of the by-products resulted from oxidation of the oil are carcinogenic which

M. J. Jalil et al., *Bio-Based Epoxides*,
SpringerBriefs in Applied Sciences and Technology,
https://doi.org/10.1007/978-981-95-5578-9_4

cause cancer [10]. Meanwhile, for the environmental effects, the improper discharge of waste cooking oil will lead to the clogging of sewer pipes and drains causing flooding or sewer overflows [11]. The inappropriate discharge of waste cooking oil into the water system will change the oxygenation process which will destroy aquatic life by covering the surface of the water and preventing oxygen from dissolving [12]. Waste cooking oil has great potential to be commercialized as it can be used in the production of products such as biodiesel, polyurethane (polyol), and bitumen which can reduce the dependency on natural resources [13]. However, there is a lack of study of waste cooking oil, especially palm oil based, as feedstock for epoxidation.

Beside homogeneous catalyst, there is another one of the epoxidation process methods which involves the use of heterogeneous catalysts, such as zeolite, titanium or tungsten [14]. In situ epoxidation refers to a process in which the epoxide is formed within the reaction mixture and not isolated as a separate product [15]. This type of epoxidation can be performed using a heterogeneous catalyst, which is a catalyst that is insoluble in the reaction mixture and forms a separate phase [16]. The peracid oxidant is produced in situ when carboxylic acid reacts with hydrogen peroxide in the presence of a mineral acid as the catalyst. However, the presence of peracid should be avoided during the epoxidation process because of their disadvantages. Additionally, there are a number of problems with heterogeneous catalyzed epoxidation that need to be improved. These include the selectivity to epoxidized is relatively slow due to oxirane ring opening taking longer time and additional separation process of acidic byproduct requires further apparatus and equipment.

4.2 Epoxidation of Waste Cooking Oil

In a 500-mL three-neck round-bottom flask fitted with a magnetic stirrer, a thermometer, and a reflux condenser, 100 g of WCO was first mixed with formic acid. In a thermostatic oil bath, the flask was immersed. The mixture was heated to the desired temperature after being stirred at a certain rpm. Following that, catalyst (%) was added to the mixture, followed by the addition of hydrogen peroxide drop by drop. The reaction was run for 6 h, after which 5 mL of sample was withdrawn with a syringe for oxirane material analysis using the AOCS Official Method Cd-957 [17].

4.3 Determination of Relative Conversion to Oxirane (%RCO)

The experimental oxirane oxygen content (OOCexp) was determined using the AOCS Tentative Method Cd 9–57 [18]. The OOC analysis was carried out to study the effect of the process parameters on the overall reaction progress, determination the yield of EWCO-PO, and to provide experimental data in the fitting procedure of kinetic modeling. The OOC indicates the amount of oxygen contained in oxirane (or epoxy) cyclic functional group. The OOC value obtained from experiment is termed the experimental oxirane oxygen content (OOCexp). The value of OOCexp is a good indicator to determine that the reaction has completed [19].

The experimental procedure of obtaining the value of OOCexp can be briefly explained as follows: 10 mL of acetic acid was added into sample weighed in an Erlenmeyer flask. The mixture was shaken rigorously. Next, two drops of crystal violet indicator were dropped into the mixture. The mixture was then titrated while stirring with hydrogen bromide (HBr) until the mixture became bluish green in color and the color remained for at least 30 s. The experimental oxirane oxygen content (OOCexp), in moles/100 g is given by Eq. 4.1.

$$\mathrm{OOC}_{\mathrm{experimental}} = 1.6 \times N \times \frac{(V - B)}{W} \tag{4.1}$$

where N is the normality of HBr, V is the volume of HBr solution used for the blank (in mL), B is the volume of the HBr solution used for titration (in mL), and W is the weight of the sample (in g). The IV is used to calculate the theoretical oxirane oxygen content (OOCtheo) using the Eq. 4.2.

$$\mathrm{OOC}_{\mathrm{theoretical}} = \left\{\left(\frac{X_0}{A_i}\right) \Big/ \left[100 + \left(\frac{X_0}{2A_i}\right)(A_0)\right]\right\} \times A_0 \times 100 \tag{4.2}$$

where IV_0 is the initial IV, A_0 is the molar mass of oxygen, and A_i is the molar mass of iodine. Lastly, the percentage of relative conversion to oxirane (RCO) is calculated using the Eq. 4.3. RCO is a measure of the efficiency of epoxidation reaction, in converting a starting material into the desired product, which is the oxirane [20]. The RCO is calculated by dividing the amount of oxirane produced in the reaction by the amount of starting material used and expressing the result as a percentage. In addition, the RCO is often used as an indicator of the quality of the final product, with higher RCO values indicating a more efficient and complete reaction.

$$\mathrm{RCO} = \frac{\mathrm{OOC}_{\mathrm{experimental}}}{\mathrm{OOC}_{\mathrm{theoretical}}} \times 100 \tag{4.3}$$

4.4 Kinetic Modeling of Epoxidation Waste Cooking Oil

The epoxidation of oleic acid was kinetically modeled by determining the reaction rate equation. This involved solving the numerical rate equation and computing the error between the simulated and experimental values [14]. The kinetic constants of k_1, k_2, and k_3 are shown in Eqs. 4.4 and 4.5. The kinetic data were represented by a model assuming constant volume in each phase throughout the epoxidation process, with all reactions considered homogeneous.

$$AA + HP \underset{k2}{\overset{k1}{\rightleftarrows}} PRA + H_2O \tag{4.4}$$

$$PRA + WCO \rightarrow^{k3} EWO + AA \tag{4.5}$$

AA, HP, PRA, WC, and EWO are acetic acid, hydrogen peroxide, peracetic acid, waste cooking oil, and epoxidized waste cooking oil, respectively. Based on rate constants k_1, k_2, and k_3 kinetic model for the epoxidation process and epoxide ring degradation can be formulated using the set of simultaneous differential equations given in Eqs. 4.6 to 4.11.

$$\frac{d[AA]}{dt} = -k_1[AA][HP] + k_2[PRA][Water] + k_2[PRA][WCO] \tag{4.6}$$

$$\frac{d[HP]}{dt} = -k_1[AA][HP] + k_2[PRA][Water] \tag{4.7}$$

$$\frac{d[PRA]}{dt} = +k_1[AA][HP] - k_2[PRA][Water] - k_3[PRA][WCO] \tag{4.8}$$

$$\frac{d[Water]}{dt} = +k_1[AA][HP] - k_2[PRA][Water]] \tag{4.9}$$

$$\frac{d[WCO]}{dt} = -k_3[PRA][WCO] \tag{4.10}$$

$$\frac{d[EWCO]}{dt} = +k_2[PRA][WCO] \tag{4.11}$$

Equation 4.12 shows the objective function used in this kinetic model.

$$obj = \sum_{i=1}^{n} \frac{\left|EWO_i^{sim} - EWO_i^{exp}\right|}{n} \tag{4.12}$$

4.5 Effect of Temperature

The effect of temperature on the epoxidation of waste cooking oil using TiO_2 catalyst shows a significant influence on the relative conversion to oxirane (RCO%) as shown in Fig. 4.1. At 65 °C, the reaction achieves a moderate RCO% with a steady increase up to 50 min, indicating a slower but more controlled epoxidation process. At 75 °C, the conversion increases more rapidly and reaches a peak at around 40 min, followed

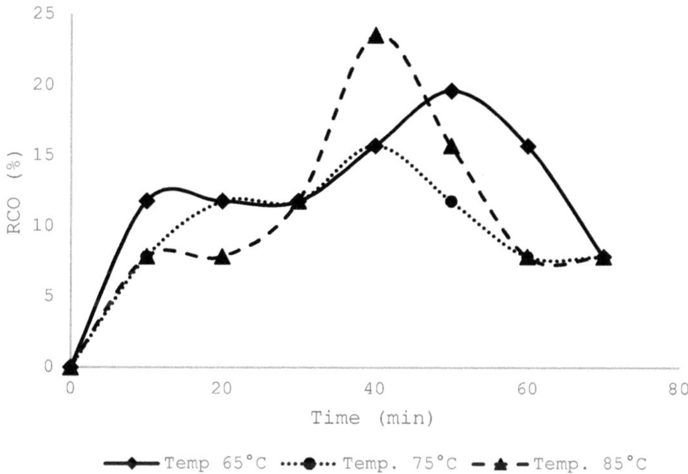

Fig. 4.1 Relative conversion to oxirane versus reaction time at different temperatures (for the epoxidation of waste cooking oil)

by a gradual decline, suggesting enhanced reaction kinetics but also potential ring-opening side reactions or degradation of the oxirane ring at longer durations. This implies that moderate temperatures allow better control over the epoxidation process [21].

In contrast, the highest temperature (85 °C) resulted in the fastest initial rise in RCO%, peaking sharply at 40 min, but quickly dropping thereafter. This trend highlights that while high temperatures accelerate the epoxidation rate, they may also promote side reactions such as epoxide ring opening or thermal degradation, leading to reduced oxirane content at longer reaction times. Overall, the optimal temperature for achieving high RCO% while minimizing unwanted side reactions appears to be 75 °C, balancing reaction efficiency and product stability.

4.6 Effect of Catalyst Loading

The influence of catalyst loading on the epoxidation of waste cooking oil using TiO_2 is evident from the comparison between 0.25 and 0.5 g catalyst doses as shown in Fig. 4.2. Interestingly, the lower catalyst amount (0.25 g) resulted in a consistently higher RCO% throughout the reaction time, reaching a peak of approximately 0.6% at 50 min. This suggests that at this loading, the catalyst was more effective in promoting oxirane ring formation without triggering excessive side reactions [22].

Conversely, the higher catalyst loading (0.5 g) showed a slower RCO% increase and a lower peak value (~0.5%), possibly due to catalyst aggregation or excessive catalytic sites accelerating side reactions such as ring opening. The plateau and

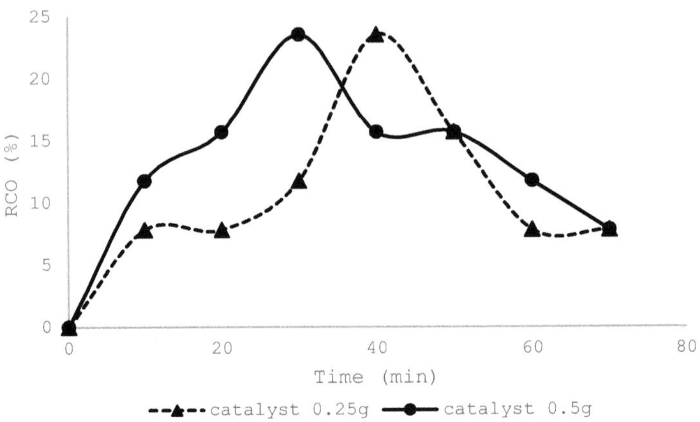

Fig. 4.2 Relative conversion to oxirane versus reaction time for the epoxidation of waste cooking oil at different catalyst loadings

eventual decline in RCO% at higher catalyst loading may also indicate product instability or oxidative degradation. These results imply that an optimum catalyst loading is essential to balance epoxide formation and stability, with 0.25 g providing more favorable conditions in this system.

4.7 FTIR Analysis of Epoxidized Waste Cooking Oil Catalyzed

The FTIR spectra of the waste cooking oil and the epoxidized product confirm successful epoxidation via the hybrid peracid mechanism in the presence of titanium dioxide as catalyst as shown in Fig. 4.3. The disappearance or significant reduction of the absorption band around 3010 cm^{-1}, attributed to the = C–H stretching of alkene groups, indicates the consumption of double bonds. Moreover, the appearance and enhancement of the peak around 820–850 cm^{-1}, corresponding to the characteristic C–O stretching of the oxirane ring, further supports the formation of epoxide. Other notable changes include the attenuation of the peak near 1740 cm^{-1}, associated with C = O stretching, suggesting partial ester transformation during the reaction. These spectral modifications collectively validate the successful conversion of unsaturated fatty acids in waste cooking oil to epoxides under the catalytic influence of titanium dioxide.

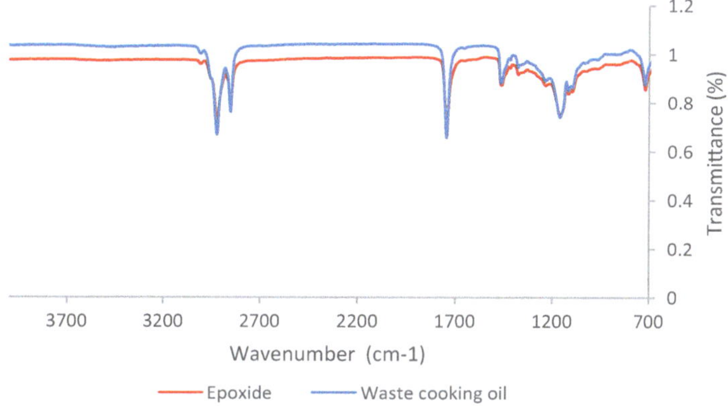

Fig. 4.3 FTIR spectra of raw waste cooking oil and epoxidized oil

4.8 Kinetic Model

In this study, simulated annealing was employed to determine the reaction rate constants for the epoxidation of waste cooking oil via a hybrid peracids mechanism in the presence of titanium dioxide as a catalyst. This optimization method is particularly effective for complex systems with large search spaces and multiple local minima, enabling the identification of near-optimal kinetic parameters, as summarized in Table 4.1. The rate constant k_2 was observed to be significantly higher than k_1, indicating that the reaction step associated with k_2 proceeds at a much faster rate compared to the step governed by k_1. This trend is consistent with previously reported findings, thereby validating the reliability of the present experimental results against established literature. Understanding these kinetic behaviors provides critical insights for process optimization, allowing for the fine-tuning of reaction parameters—such as temperature and oxidant composition to maximize epoxide yield from waste cooking oil while maintaining the stability of the oxirane ring.

The experimental and simulated concentrations of epoxidized waste cooking oil (EPWCO) exhibited good overall agreement, as illustrated in Fig. 4.4. The deviation between experimental and simulated values was approximately 0.10 mg L^{-1}, indicating that the simulation model provided reasonable accuracy in predicting the concentration profile. A notable difference, however, was observed in the time

Table 4.1 Estimated reaction rate constants from simulation

Reaction rate constant	Value (mol/L·min)
k_1	6.2
k_2	0.2
k_3	0.01
Correlation coefficient, $r = 0.9$	

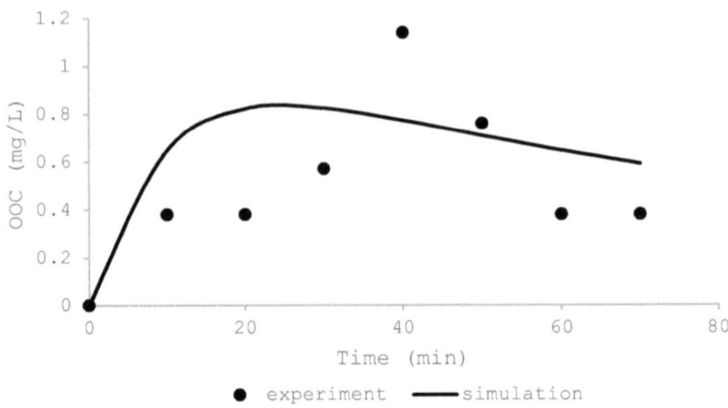

Fig. 4.4 Comparison of the oxirane content between simulation and experiment

required to reach the maximum EPWCO concentration. The simulation predicted the maximum concentration between 10 and 20 min, whereas in the experimental data, determining this exact point was challenging. This difficulty arises from the simultaneous occurrence of multiple reactions in the hybrid peracids system, which can influence the oxirane content at different rates. The simulation was based on idealized conditions, neglecting factors such as heat losses and limitations in heat transfer. In contrast, the experimental setup was subject to real-world conditions where these non-ideal effects significantly impacted the reaction dynamics.

4.9 Conclusion

The experimental results confirm that reaction temperature and catalyst loading are critical parameters in the epoxidation of WCO using TiO_2. A temperature of 75 °C provided the most favorable condition, achieving high oxirane formation with minimal ring-opening or degradation. Lower catalyst loading (0.25 g) also gave better conversion, possibly due to more controlled reactivity and less catalytic overactivity. Overall, TiO_2 proved to be an effective heterogeneous catalyst for the epoxidation of WCO, offering a promising pathway for converting waste oil into valuable bio-based materials.

References

1. A. Badan, T. Majka, The influence of vegetable-oil based polyols on physico-mechanical and thermal properties of polyurethane foams, 4763 (2017). https://doi.org/10.3390/ecsoc-21-04763

2. N.S. Hasmaruddin, M. Tahir, Synthesis of Waste Cooking Oil-Based Polyol via One-Pot Epoxidation and Hydroxylation Reaction, pp. 382–385 (2019)
3. N.M. Noor, T.N.M.T. Ismail, M.A.M. Noor, K.D.P. Palam, M.N. Sattar, N. 'Ain Hanzah, S. Adnan, Y.S. Kian, Physicochemical properties of palm olein-based polyols prepared using homogeneous and heterogeneous catalysts. J. Oil Palm Res. **34**, 167–176 (2022). https://doi.org/10.21894/jopr.2021.0034
4. M.J. Jalil, M.H.N. Ishak, I.M. Rasib, M.Z.A. Kadir, M. Noorfazlida, I.S. Azmi, Epoxidation of hybrid oleic acid-derived palm oil using an in situ performic acid mechanism **73**, 477–481 (2024)
5. H.H. Habri, A.S.A. Shahrizan, I.S. Azmi, N. Hambali, A. Shamjuddin, S. Salaeh, M.J. Jalil, Degradation autocatalytic epoxidation of oleic acid derived from palm oil via in situ performic acid mechanism. Environ. Prog. Sustain. Energy (2024). https://doi.org/10.1002/ep.14498
6. L.K. Jia, L.X. Gong, W.J. Ji, C.Y. Kan, Synthesis of vegetable oil based polyol with cottonseed oil and sorbitol derived from natural source. Chinese Chem. Lett. **22**, 1289–1292 (2011). https://doi.org/10.1016/j.cclet.2011.05.043
7. O. Study, V.B. Borugadda, V.V Goud, Synthesis of Waste Cooking Oil Epoxide as a Bio-Lubricant Base Stock: Synthesis of Waste Cooking Oil Epoxide as a Bio-Lubricant Base Stock: Characterization and Optimization Study (2014). https://doi.org/10.1166/jbeb.2014.1077
8. G.P. Maniam, H.S. Lim, N. Mat Hussin, Effect of free fatty acid on transesterification of waste cooking oil. Curr. Sci. Technol. **3**, 57–63 (2023). https://doi.org/10.15282/cst.v3i1.10290
9. P.T. Wai, P. Jiang, Y. Shen, P. Zhang, Q. Gu, Y. Leng, Catalytic developments in the epoxidation of vegetable oils and the analysis methods of epoxidized products. RSC Adv. **9**, 38119–38136 (2019). https://doi.org/10.1039/c9ra05943a
10. M.A. Rahman, N.M. Mubarak, I.S. Azmi, M.J. Jalil, Sustainable approach for catalytic green epoxidation of oleic acid with applied ion exchange resin. Sci. Rep. **13**, 1–8 (2023). https://doi.org/10.1038/s41598-023-42879-4
11. I.S. Azmi, D.N.A. Raofuddin, M.M.M. Sharif, M.J. Jalil, Epoxidation of waste cooking oil by using homogeneous catalyst. Proc. 1st Int. Conf. Chem. Sci. Eng. Technol. **2703**, 050005 (2023). https://doi.org/10.1063/5.0115008
12. F. Marriam, A. Irshad, I. Umer, M. Arslan, Vegetable oils as bio-based precursors for epoxies. Sustain. Chem. Pharm. **31**, 100935 (2023). https://doi.org/10.1016/j.scp.2022.100935
13. M.Z. Arniza, S.S. Hoong, Z. Idris, S.K. Yeong, H.A. Hassan, A.K. Din, Y.M. Choo, Synthesis of transesterified palm olein-based Polyol and rigid polyurethanes from this polyol. J. Am. Oil Chem. Soc. **92**, 243–255 (2015). https://doi.org/10.1007/s11746-015-2592-9
14. K. Polaczek, M. Kurańska, M. Auguścik-Królikowska, A. Prociak, J. Ryszkowska, Open-cell polyurethane foams of very low density modified with various palm oil-based bio-polyols in accordance with cleaner production. J. Clean. Prod. **290** (2021). https://doi.org/10.1016/j.jclepro.2021.125875
15. J. Thomas, R. Patil, Enabling green manufacture of polymer products via vegetable oil epoxides. Ind. Eng. Chem. Res. **62**, 1725–1735 (2023). https://doi.org/10.1021/acs.iecr.2c03867
16. E. Vanags, M. Kirpluks, U. Cabulis, Z. Walterova, Highly functional polyol synthesis from epoxidized tall oil fatty acids 1–8 (2018)
17. T.N.M. Tuan Ismail, N.A. Ibrahim, M.A. Mohd Noor, S.S. Hoong, K.D. Poo Palam, S.K. Yeong, Z. Idris, C.M. Schiffman, I. Sendijarevic, E. Abd Malek, N. Zainuddin, V. Sendijarevic, Oligomeric composition of polyols from fatty acid methyl ester: the effect of ring-opening reactants of epoxide groups. J. Am. Oil Chem. Soc. **95**, 509–523 (2018). https://doi.org/10.1002/aocs.12044
18. S.S. Narine, J. Yue, X. Kong, Production of polyols from canola oil and their chemical identification and physical properties. J. Am. Oil Chem. Soc. **84**, 173–179 (2007). https://doi.org/10.1007/s11746-006-1021-5
19. S. Meadows, M. Hosur, Y. Celikbag, S. Jeelani, Comparative analysis on the epoxidation of soybean oil using formic and acetic acids **26**, 289–298 (2018)
20. N.H. Rahim, M.J. Jalil, N.M. Mubarak, I.S. Azmi, G. Anbuchezhiyan, Catalytic epoxidation of unsaturated fatty acids in palm stearin via in situ peracetic acids mechanism. Sci. Rep. **15**, 1–12 (2025). https://doi.org/10.1038/s41598-025-89399-x

21. P.D. Meshram, H.V. Patil, R.G. Puri, Producing phosphate-polyols by ring-opening hydrolysis of wild safflower oil epoxides **10**, 206–215 (2017)
22. N.D. Kasmin, I.S. Azmi, S.D. Nurherdiana, F.A.M. Yusof, M.J. Jalil, Chemical modification of linoleic acid via catalytic epoxidation of corn oil: A sustainable approach. Environ. Prog. Sustain. Energy (2024). https://doi.org/10.1002/ep.14362

Chapter 5
Extraction of Rubber Seed Oil as a Feedstock for Epoxidation Process via Mixed Solvent System: Preserving Unsaturation Content of Oil

5.1 Vegetable Oils and Unsaturation Content Determination

As a replacement to fossil sources, vegetable oils have gained a lot of interest as prospective feedstocks for the production of renewable energy. Even though vegetable oils are abundant in nature, they are not highly reactive. However, chemical changes, particularly those that occur at their double bonds, have the potential to increase their reactivity and broaden the range of commercial applications for vegetable oils. Chemical changes such as epoxidation and hydroxylation are reactions that are the most frequently studied [1]. Edible vegetable oils, such as sunflower oil, rice bran oil, and palm oil in non-edible applications have raised significant concerns on the food sustainability and security of food. Rubber seed oil, jatropha oil, karanja oil, and neem oil are the examples of non-edible oils that have been extensively studied as potential substitutes for many industrial applications. The epoxidation of non-edible oils has been the focus of recent efforts to generate valuable intermediates for a variety of industries. Numerous studies have been conducted to examine the epoxidation of non-edible oils such as jatropha oil [2], rubber seed oil [3], and melon seed oil [4]. The potential of these feedstocks to be environmentally friendly and sustainable has been the subject of a significant amount of study. The shift toward the use of oils that are not edible not only addresses issues regarding the long-term viability of food, but it also paves the way for the development of novel uses in the sectors of industrial materials and renewable energy.

The unsaturation level of vegetable oils is a critical factor to the epoxidation process, due to the reaction preferentially targeting the carbon–carbon double bonds ($C = C$) in the fatty acid chains of triglycerides [5]. Epoxidation cannot occur in the absence of these double bonds. Epoxidation involves the incorporation of an oxygen atom over a double bond, resulting in the formation of a three-membered cyclic ether known as an oxirane ring or epoxy group. An oil with a higher number of double bonds possesses more potential sites for epoxidation. Consequently, oils abundant

M. J. Jalil et al., *Bio-Based Epoxides*,
SpringerBriefs in Applied Sciences and Technology,
https://doi.org/10.1007/978-981-95-5578-9_5

in unsaturated fatty acids, especially polyunsaturated fatty acids which contain more than two double bonds and monounsaturated fatty acids which contain one double bond, serve as optimal feedstocks for epoxidation [6]. A variety of approaches exist to detect and quantify vegetable oil unsaturation content. The most conventional technique is Iodine Value (IV) determination by titration. This method quantifies the iodine absorption of the oil, indicating the quantity of double bonds present. Iodine interacts with C = C bonds; therefore, an increase in iodine consumption (measured in centigrams per gram) correlates with increasing unsaturation. The Wijs method is a prevalent titrimetric technique employed for the determination of iodine value. The process necessitates reagents such as iodine monochloride and potassium iodide, and entails titrating the oil sample to determine the iodine number as an indicator of unsaturation [7]. Raman spectroscopy facilitates the rapid and non-invasive assessment of unsaturation by examining vibrational bands characteristic of C = C stretching (~1661 cm^{-1}) and CH_2 deformation (~1444 cm^{-1}). The intensity ratio of these bands quantitatively correlates with the level of unsaturation. This technology is applicable for process monitoring owing to rapid measurement times and little sample preparation [8]. Additional chromatographic techniques encompass gas chromatography (GC) which, when integrated with mass spectrometry (MS), can provide elevated precision, molecular specificity, and sensitivity. This technique entails the conversion of fatty acids into methyl esters, succeeded by capillary gas–liquid chromatography to isolate and quantify certain unsaturated fatty acids, including omega-3 and omega-6. This approach offers comprehensive profiling of fatty acid composition and unsaturation levels; however, it is more intricate and time-consuming, making it appropriate for in-depth compositional investigation rather than standard IV testing [9]. Apart from that, thin-layer chromatography (TLC) is a conventional technique. Nevertheless, it cannot provide precise measurement, and exposure of the TLC plate to ambient oxygen may result in lipid oxidation [10].

5.2 The Importance of Unsaturation Content Toward Epoxidation Process

Due to the fact that the presence and number of carbon–carbon double bonds (C = C) in the fatty acids directly impact the epoxidation yield and efficiency, the unsaturation content of vegetable oils is an extremely important factor in the epoxidation process. Through the addition of an oxygen atom, the reaction takes place in which the C = C bonds that are present in unsaturated fatty acids are transformed into three-membered oxirane rings, which are known as epoxides. Oils that have a higher unsaturation level (contain more double bonds per molecule) offer more reactive sites for epoxidation, which ultimately results in a higher epoxide concentration [11]. Unsaturation level of vegetable oils is related to the number of epoxide groups produced. Oils that are abundant in monounsaturated, diunsaturated, and polyunsaturated fatty acids, including oleic (C18:1), linoleic (C18:2), and α-linolenic (C18:3) acids, exhibit distinctively different levels of epoxidation efficiency. For example, α-linolenic acid, which possesses three double bonds, has the ability to generate tri-epoxides, which enhances the functionalization of the compound [6].

The characteristics of oil are altered by epoxidation. The incorporation of epoxide rings results in an increase in the polarity and stability of oil, which in turn improves its compatibility with polymers and makes it possible to manufacture important intermediates such as plasticizers, stabilizers, lubricants, and polyols [12]. The findings from the Thin Layer Chromatography (TLC) examination of some epoxidized vegetable oils indicated that the more polar compounds exhibited prolonged adsorption on the polar silica-coated aluminum stationary phase. The interaction of a carbon–carbon double bond with reactive oxygen results in the incorporation of an oxygen atom, transforming the original double bond into a three-membered epoxide, or oxirane ring. Consequently, the oxirane ring epoxides exhibit significant reactivity for the production of many molecules, including alcohols (polyols), glycols, olefinic compounds, lubricants, plasticizers, and polymer stabilizers [13].

With the help of percarboxylic acids, the Prileschajew reaction is typically utilized in order to carry out the epoxidation of unsaturated fatty acid molecules. Nevertheless, this technique, which makes use of powerful mineral acids as catalysts for the in-situ generation of peracids, is plagued by a number of drawbacks. These include the low selectivity for epoxide formation, which is caused by the opening of the oxirane ring in an acidic medium, the corrosive nature of acids, and the unstable nature of peracids [14]. Conversely, direct enzymatic procedures present a solution for the more selective epoxidation of unsaturated fatty acids. Enzymatic epoxidation employing fungal peroxygenases exhibits selective epoxidation of unsaturated fatty acids while leaving saturated fatty acids largely untouched. This is because unsaturation has an effect on enzyme-catalyzed epoxidation [15]. Different enzymes have preferences for particular unsaturation patterns, which can have an effect on the selectivity and yield of epoxidation. In a study employing three different enzymes, the epoxidation of fatty acids favors substrates with higher unsaturation levels, particularly oleic acid, linoleic acid, and α-linolenic acid to produce diepoxides and triepoxides [6]. In conclusion, the higher and more varied the unsaturation levels in vegetable oils, the greater the possibility for efficient and high-yield epoxidation. This is essential for the production of epoxidized vegetable oils that have improved functional qualities for use in industrial applications.

5.3 Effect of Solvent Polarity Toward Oil Yield and Unsaturation Content

In response to the growing need for bio-based materials made from the epoxidized vegetable oil, the extraction of vegetable oil has emerged as an important topic of research. Two prevalent techniques for extracting oil from vegetable seeds are solvent extraction and mechanical pressing. In the mechanical pressing technique, temperature control and pressure were employed to extract seed oil, but the oil recovery from the vegetable seeds is relatively low [16]. On the other hand, solvent extraction is a technology that is highly utilized for the purpose of extracting vegetable oil. This method is not only simple to scale up, but it also yields a high amount of oil, is simple, and requires less energy. Extensive study is conducted utilizing the Soxhlet apparatus for the extraction process in order to optimize oil yield recovery. Some of the major aspects that are investigated include the polarity of the solvent used, the extraction time, the ratio of solids to solvents, the seed size, and the operating temperature. The Soxhlet apparatus works by applying both the percolation and the reflux extraction processes, which constantly extract the oil with a new solvent through the use of the ideas of syphoning and reflux. The extraction rate is higher than that of maceration or percolation, and it requires less time and solvent than the other two methods [17].

Solvent extraction remains the primary method for recovering vegetable oil from a variety of seeds. Despite the fact that the Soxhlet extraction process recovers a high oil yield, the solvents that are used are extremely important. This is due to the fact that the majority of solvents are hazardous upon lengthy periods of exposure. Additionally, the solubility and selectivity of the solvents have an impact on the quality of the vegetable oil. The oil yield is affected by the polarity of the solvents, which is determined by the polarity of the solute that is being sought for. Solvents that have polarity values that are quite near to those of the solute also have a high level of efficiency, and vice versa [18]. The solvent type, whether a single solvent or a mixed solvent system, can directly affect both the yield and quality of the extracted oil. A key quality measure is the degree of unsaturation, usually expressed as the ratio or percentage of monounsaturated (MUFA) and polyunsaturated fatty acids (PUFA) in the oil. Comprehending the impact of solvent selection on unsaturation levels aids in optimizing processes for nutritional value, shelf stability, and regulatory adherence.

5.4 Single Solvent

In this study, solvent extraction of rubber seed oil was performed using four different solvents (ethanol, methanol, petroleum ether, and hexane) having different polarity at 8 h of extraction time. Kernel size and solid-to-solvent ratio were fixed at 0.5 mm and 0.07 g/ml, respectively. Solvent's polarity affected the oil yield and quality of the extracted oils, as the quantity of fatty acids in vegetable oil differed significantly across different extraction solvents and methods [19]. Hexane, a non-polar solvent, is typically utilized due to its ability to produce high seed oil yield [20–23]. Another non-polar solvent, petroleum ether, serves as an alternative to n-hexane and has been extensively investigated with the ability to extract vegetable oils with higher oil yields compared to n-hexane [24]. Alcohols such as ethanol and methanol are polar solvents. Based on Fig. 5.1, hexane and petroleum ether consistently produced higher oil yield (57.54% and 55.94%, respectively) at 8 h of extraction time compared to polar solvents, ethanol and methanol (54.19% and 51.68%). Among all solvents, hexane showed the highest oil yield, confirming hexane as a widely used solvent in vegetable extraction. Polar solvents, on the other hand, produced the lowest oil yield, which was attributed to the existence of fewer fatty acids with polar long chains of hydrocarbon in rubber seeds than non-polar ones [25]. This suggests that polarity mismatch limits oil solubility despite prolonged extraction. Polar solvents, such as alcohols, are recognized for their ability to extract elevated quantities of free fatty acids (FFAs) and undesirable substances, including proteins, carbohydrates, Maillard reaction products, phosphatides, and various other chemicals [26]. Overall, the performance of the solvents can be ranked as: hexane > petroleum ether > ethanol > methanol.

Polarity of solvent can be technically measured by three parameters which are relative polarity, polarity index, and dielectric constant. Relative polarity is a method for quantifying the polarity of a molecule or liquid. It demonstrates how the electric charge is distributed throughout the molecule, resulting in a dipole moment in which one half is partially positive, and the other is largely negative. Relative polarity indicates how polar a molecule is in comparison to water. Polarity index, on the other hand, is a number that indicates how well a solvent interacts with solutes, based on its polarity. It classifies solvents according to their polarity. A higher polarity index value indicates that the solvent is more polar and interacts with the solute. The dielectric constant, also known as relative permittivity, indicates how efficiently a material retains electrical energy in an electric field. In solvents, it determines how much a solvent may reduce the electric field between charged particles, indicating the

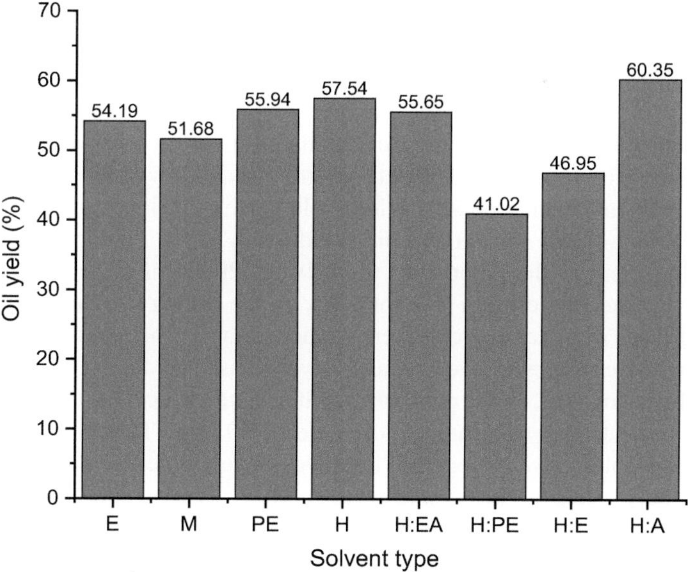

Fig. 5.1 The effect of polar and non-polar solvents toward rubber seed oil yield (E: Ethanol, M: Methanol, PE: Petroleum Ether, H: Hexane, EA: Ethyl acetate, A: Acetone)

Table 5.1 Relative polarity, polarity index, and dielectric constant values for selected solvents [27]

Solvent type	Relative polarity	Polarity index	Dieletric constant
Ethanol	0.654	5.2	30.0
Methanol	0.762	5.1	32.7
Acetone	0.355	5.1	20.7
Ethyl acetate	0.228	4.4	6.02
Petroleum ether	0.117	0.1	1.90
Hexane	0.009	0	1.88

solvent's polarity on a wider scale. Solvents possessing a dielectric constant below 15 are typically classified as non-polar while solvents with values more than 15 are called polar [27]. Instead of focusing on solvent polarity, this study highlighted the impact of solvent dielectric constant toward its ability to increase rubber seed oil yield and quality. By comprehending and manipulating the dielectric constant, researchers can improve the efficiency and selectivity of fatty acid extraction, especially in the context of vegetable oil extraction. Based on Table 5.1, hexane, a non-polar solvent with lowest polarity index (0) and dielectric constant (1.88), showed the ability to extract highest oils yield among all solvents, particularly nonpolar lipids.

5.5 Mixed Solvents

Complex oils such as vegetable oils have both polar and nonpolar fatty acids. Medium-polarity solvents can extract a broader range of fatty acids from vegetable oil seeds [28]. Recently, the research that is being conducted to preserve the unsaturation level of vegetable oil is focusing on the usage of two-solvent combinations or mixed solvents. In a mixed-solvent system, a secondary solvent, often referred to as a co-solvent, is incorporated into a sole solvent. This results in an increase in the solubility of the solvent system, a reduction in the toxicity of the solvent, and an adjustment in the polarity of the solvent [29]. This is particularly advantageous when the solute has a low solubility in the primary solvent alone. In this study, four types of mixed solvents were studied in the extraction of rubber seed oil. The performance of those solvents was analyzed based on the oil yield percentage obtained. Based on Fig. 5.1, the combination of hexane and acetone (H:A), a nonpolar and polar solvent, yields the highest oil yield of 60.35%. This means that mixing polar and nonpolar solvents together improves extraction efficiency. The mixtures of hexane, ethyl acetate (H:EA), and acetone (H:A) are most likely polar-nonpolar. The latter performs better, indicating that the balance of solvent polarity in mixes is critical. Hexane plus ethanol (H:E) creates 46.95% less oil than pure hexane. This demonstrates how the polarity and compatibility of the solvents in a mixture are also important. The combination of petroleum ether and ethanol (P:E) generates the least

(41.02%), indicating that blending these two solvents may not function as well as planned. This could be due to poor solvent–solute interaction or dissolution.

5.6 Fatty Acid Compositions of Rubber Seed Oil

Rubber seed oil obtained from the solvent extraction using single and mixed solvents was analyzed for their fatty acid compositions. Table 5.2 shows how various solvents, including ethanol, methanol, petroleum ether, hexane, and hexane:acetone, altered the fatty acid composition of rubber seed oil (RSO). While the total amount of unsaturated fatty acids (UFAs) was consistently high in all solvents (more than 90%), the profiles of individual fatty acids varied slightly, indicating how the solvent interacted with the lipids and how selective the extraction was. Gamma-linolenic acid accounted for the majority of the fatty acids identified in all solvents, ranging from 39.6% (hexane) to 42.7% (methanol). The slightly improved recovery with methanol indicates that polar solvents may aid in the dissolution of polyunsaturated fatty acids (PUFAs) such as GLA by interacting well with their polarizable double bonds. However, petroleum ether and hexane:acetone removed similar amounts (~41.5–41.6%), indicating that GLA can be extracted using any solvent polarity. This consistency shows that rubber seed oil naturally has a lot of UFAs and that the kind of solvent doesn't change the level of unsaturation too much. The solvent does, however, change the distribution of certain UFAs (GLA, linolelaidic, and elaidic), which shows that they are not all the same.

In other study, oils extracted with hexane often preserve their original fatty acid compositions with negligible effects on unsaturation levels. Only negligible (1–3%) variations in unsaturated fatty acid concentration have been noted in comparison to other solvents [30]. Apart from hexane, alcohols, such as ethanol, isopropanol, and methanol, have polarity that can extract more polar lipids and minor bioactive substances. These solvents are observed to have slightly higher polyunsaturated fatty acids (PUFA) content relative to hexane-extracted oil [29]. Other solvents such as ethyl acetate also shows good extraction properties and ability to preserve unsaturation level in vegetable oil similar to hexane. The fatty acid profile of pistachio oil was evaluated by using a variety of solvents, including ethanol, hexane, dichloromethane, and ethyl acetate. The range of 88.49% was determined to be the quantity of major unsaturated fatty acids utilizing the Soxhlet equipment with ethyl acetate [31]. Table 5.3 summarizes different types of solvents that managed to preserve the highest unsaturation content of fatty acids in solvent extraction process of rubber seed oil.

In a study employing mixed solvent systems in the extraction of rice bran oil, a positive finding. In comparison to conventional solvent extraction, the peroxide value, acid value, iodine value, and fatty acid composition of rice bran oil extracted with the novel mixed solvent (isopropanol and cyclohexane) were similar to those obtained using n-hexane. The percentage of total unsaturated fatty acid (75.4%) obtained was as high as the one obtained by n-hexane despite the lower toxicity of cyclohexane

Table 5.2 Fatty acid composition of rubber seed oil obtained from different solvents

Fatty acid composition	Solvent type				Hexane:Acetone
	Ethanol	Methanol	Petroleum ether	Hexane	
Gamma.Linolenic acid	41.3	42.7	41.5	39.6	41.6
Linolenic acid	19.1	20.2	19.6	17.2	19.1
Linolelaidic acid	29.7	20.6	25.5	10.0	21.7
Elaidic acid	ND	7.4	3.9	ND	7.8
Palmitic acid	8.6	8.6	8.3	8.2	8.7
Eicosanoic acid	0.3	0.3	0.3	ND	0.3
Eicosenoic acid	0.2	0.2	0.2	ND	0.2
Myristic acid	0.1	ND	0.2	0.4	0.1
Palmitoleic acid	0.2	ND	0.2	ND	0.1
Docosanoic acid	0.4	ND	0.2	ND	0.2
Heptadecanoic acid	ND	ND	0.1	ND	0.1
Oleic acid	ND	ND	ND	23.7	ND
Stearic acid	ND	ND	ND	10.2	ND
Dodecanoic acid	ND	ND	ND	0.7	ND
UFA	90.1	90.9	90.5	90.5	90.2

Table 5.3 Percentage of unsaturated fatty acid of various vegetable oils obtained using various solvents

Solvent	Vegetable oil	Method of extraction	Percentage of unsaturation content	References
Hexane	Rubber seed oil	Solvent extraction	80.5	[30]
Hexane	Rubber seed oil	Solvent extraction	80.09	[25]
Acetone	Safflower seed oil	Solvent extraction	90.77	[19]
Ethyl acetate	Pistachio oil	Solvent extraction	88.493	[31]
Hexane	Flaxseed oil	Solvent extraction	85.29	[32]

and isopropanol [33]. In another study employing a co-solvency approach in the extraction of rubber seed oil (RSO) using hexane–acetone, a higher variety of fatty acid composition was obtained compared to conventional sole hexane [30]. The percentage of unsaturated fatty acid obtained was higher (82.4%) compared to oil extracted using a single solvent (80.5%). This finding showed the efficacy of the innovative mixed solvent to preserve unsaturated fatty acids in rubber seed oil. The addition of acetone to hexane improves the solubility and adjusts the overall polarity of the solvent system to improve the extraction process, which directly preserves unsaturated fatty acids from rubber seed. The addition of acetone showed a good strategy, as another study has proved that acetone in the extraction of safflower oil yielded the highest concentrations of unsaturated fatty acids of 90.77% [19]. Apart from vegetable oil extraction, mixed solvent systems also worked for microalgae oil extraction, where a higher iodine value of 108.48 (g I_2/100 g), indicative of high unsaturation content, was obtained using mixed solvents of ethanol and hexane compared with sole hexane (102.15 g I_2/100 g) and sole ethanol (100.91 g I_2/100 g) [34].

5.7 Other Extraction Parameters

5.7.1 Lower Temperature Extraction

Unsaturated fatty acids, particularly polyunsaturated fatty acids (PUFAs) such as linoleic and alpha-linolenic acids, are markedly prone to heat breakdown [35]. Upon exposure to elevated temperatures, they may undergo several adverse reactions, including oxidation and polymerization. Heat stimulates the interaction of double bonds with atmospheric oxygen, resulting in the creation of free radicals, hydroperoxides, and ultimately secondary oxidation products such as aldehydes, ketones, and polymers. These interactions decompose unsaturated fatty acids, diminishing their concentration and resulting in rancidity and undesirable flavors [36]. Apart from that, unsaturated fatty acids can combine to form polymers and undergo geometric as well as positional isomerization when heated to high temperatures [37]. Cold pressing, which works at much lower temperatures (usually below 50 °C), helps to minimize heat exposure that makes the degradation reactions mentioned above. The fragile double bonds in the fatty acids stay mostly intact when they are kept away from high temperatures. This approach cannot be achieved using solvent extraction methods that usually use high temperature depending on the solvent type used, which can speed up oxidation and promote unsaturation structure breakdown or be altered if not well regulated.

5.7.2 Oxygen Exclusion Extraction

Apart from high temperature, unsaturated fatty acids which are exposed to oxygen are prone to lipid oxidation that can cause unsaturated bond degradation. Polyunsaturated fatty acids (PUFAs) are significantly more prone to oxidation due to the presence of several methylene groups situated between double bonds, referred to as bisallylic groups, which are characterized by weaker C–H bonds and so are more readily subjected to hydrogen extraction [38]. The process's mechanism consists of a rapid reaction between a chain-carrying carbon radical and oxygen, followed by a slow rate-determining propagation phase in which a peroxyl radical abstracts hydrogen from an organic substrate [39]. The propagation of peroxyl radicals can also occur through addition reactions with alkenes. H-atom transfer is the mechanism by which phenolic inhibitors of autoxidation (antioxidants) compete for peroxyl radicals. A second peroxyl is trapped by the generated aryloxy radical in the transformation, resulting in the formation of non-radical products and the termination of the chain [39]. The principal results of autoxidation are peroxides or hydroperoxides; nevertheless, these molecules are often unstable and disintegrate into aldehydes, ketones, and other reactive entities [39].

In order to minimize lipid oxidation, the extraction process should be performed with low exposure to oxygen. The vegetable oils should be stored in sealed containers, performing nitrogen blanketing and reduced headspace in a container. In solvent extraction process, although some heat is essential for solvent efficacy and elimination, operations should be maintained at the lowest effective temperatures during extraction, miscella filtration, and solvent removal procedures. Vacuum condition can be applied during the solvent removal process to reduce operating temperature. The activation energy for radical formation (initiation) is provided by heat, which also substantially accelerates all stages of the oxidation chain reaction, particularly hydroperoxide decomposition. The rate of oxidation can be doubled with each 10 °C increase in temperature. In comparison to hexane, the utilization of alternative green solvents such as ethanol and ethyl acetate frequently enables the oil to be extracted and evaporated at lower temperatures, thereby substantially reducing thermal stress. Another green solvent that is frequently studied for solvent extraction is supercritical carbon dioxide. This solvent has a low temperature and non-oxidative properties that have been extensively suggested as a solvent for oil extraction to preserve unsaturation content. Carbon dioxide is a secure solvent due to its non-flammable, non-toxic, and cost-effective nature, and it facilitates supercritical operation at low pressures and temperatures to avoid unsaturation degradation [40].

5.8 Assisted Extraction Techniques

5.8.1 Microwave-Assisted Extraction

One of the most recent methods for extracting essential oils, fats, and oils is microwave-assisted solvent extraction (MAE), which offers benefits over traditional solvent extraction. Among MAE's appealing features are its quick extraction speed, low solvent use, and minimal impact on thermolabile components. MAE helps preserve the unsaturation content of vegetable oils primarily by enabling faster extraction with lower thermal degradation compared to conventional methods. Reduced extraction time and lower energy consumption: MAE rapidly heats the moisture inside seed cells, causing cell rupture and efficient oil release in shorter times, which minimizes exposure of unsaturated fatty acids to heat-induced oxidation or degradation. This technique also possesses lower overall extraction temperature as MAE uses microwave energy for internal heating rather than prolonged external heating, the process temperature can be controlled more precisely, preventing thermal breakdown of sensitive unsaturated fatty acids like polyunsaturated fatty acids (PUFAs).

In a study employing MAE approach in extraction of castor oil from castor seeds obtained higher unsaturated fatty acids (99%) compared to extraction using hexane extraction method (92%) [41]. Similar finding was obtained in the MAE of *Hevea brasiliensis* seeds where MAE method produced oil with higher unsaturated fatty acid content (iodine value: 150.5 g I_2/100 g) compared to conventional solvent extraction method (iodine value: 121.9 g I_2/100 g) [42]. In the oil extraction of dragon fruit seeds using a microwave, it also possesses higher degree of unsaturation fatty acids which is determined by the iodine value. Compared to the conventional solvent extraction method, which has an iodine value of 127 ± 0.98 (g I_2/100 g), the MAE seed oil has a higher value of 131 ± 1.01 (g I_2/100 g) [43]. In summary, microwave-assisted extraction can enhance oil yield and quality while maintaining the unsaturation content by reducing extraction time, lowering thermal impact, and preserving antioxidants that protect unsaturated fatty acids. This makes MAE a promising technique for producing nutritional, stable vegetable oils with high unsaturation.

5.8.2 Ultrasonic-Assisted Extraction

Ultrasonic-assisted extraction (UAE) is a technique that utilizes either an ultrasound device or an ultrasonic bath that operates at frequencies of 20 kHz and 40 kHz, respectively. This technique is based on a concept of cavitation or oscillation. The energy transfer from voids created by vibrations which are produced by ultrasonic waves to solid particulates that are submerged in the extraction. Additionally, cavitation bubbles increase in close proximity to the solid surface and implode with a greater amplitude, which results in the rupture of the cell wall. The required compounds that are contained within the bubbles are transferred into the solvent medium in a more expedient manner as a result of this process. This is a rapid and effective extraction technique that utilizes ultrasound to accelerate the extraction process by generating rapid movement of solvents, thereby increasing the mass transfer speed [44]. UAE preserves unsaturated fatty acids in vegetable oils by facilitating extraction under milder circumstances, hence minimizing exposure to heat and oxygen that often lead to oxidation and destruction of these sensitive fatty acids. Ultrasound waves generate cavitation bubbles that disturb cellular matrices, enabling the release of oil at reduced temperatures and abbreviated extraction durations relative to traditional procedures. This gentle treatment reduces thermal degradation and oxidation of unsaturated fatty acids, which are susceptible to harm from elevated temperatures or extended exposure to air [45].

In a study of using UAE method in the extraction of raspberry seed oil, UAE was able to provide a higher content of beneficial unsaturated fatty acids (UFA) compared to conventional solvent extraction, whereas conventional Soxhlet extraction resulted in a higher amount of saturated fatty acids. Total UFA content subjected to UAE was 92.8% compared to oil subjected to solvent extraction, which was only 89.07% [46].

5.9 Conclusion

This study underscores the critical importance of preserving the unsaturation content of vegetable oils during extraction, particularly when these oils are used as feedstock for the epoxidation process. The carbon double bonds (C = C) within unsaturated fatty acids serve as essential reactive sites for the formation of epoxides; consequently, any degradation or oxidation of these bonds during the extraction process can directly diminish the efficiency and yield of the epoxidation process. This study reveals the effect of solvent polarity toward rubber seed oil yield and fatty acid compositions. The extracted rubber seed oil naturally has a lot of UFAs, and the type of solvent used doesn't change the level of unsaturation too much. The solvent does, however, change the distribution of certain UFAs (GLA, linolelaidic, and elaidic), which shows that they possessed different selectivity. Our comparative analysis of extraction conditions, solvent choices, and assisted extraction techniques reveals that elevated temperatures and exposure to prolonged oxygen substantially reduce unsaturation. Conversely, milder extraction parameters and the application of assisted technologies, such as ultrasound-assisted and microwave-assisted extraction, demonstrate considerable potential in preserving the quality of unsaturated components. These advanced methods facilitate mass transfer while concurrently minimizing exposure to oxidative and thermal degradation. Ultimately, the successful preservation of unsaturation not only guarantees a more efficient epoxidation reaction but also aligns with the broader objective of cultivating sustainable, high-performance bio-based materials.

References

1. K. Nwosu-Obieogu, F.O. Aguele, L. Chiemenem, Optimization on rubber seed oil epoxidation process parameters using response surface methodology. Iran. J. Chem. Chem. Eng. **40**(5), 1575–1583 (2021). https://doi.org/10.30492/ijcce.2020.40345
2. L.L.T. Mai, M.M. Aung, S.A.M. Saidi, P.S. H'ng, M. Rayung, A.M. Jaafar, Non edible oil-based epoxy resins from jatropha oil and their shape memory behaviors. Polymers (Basel) **13**(13), 1–12 (2021). https://doi.org/10.3390/polym13132177
3. N.T. Thuy, P.N. Lan, Epoxidation of vietnam rubber seed oil by using peroxophosphatotungstate catalyst complex based on Na2WO4/H3PO4/H2O2 with the presence of phase-transfer catalyst. Mol. Catal. **509**(April), 111645 (2021). https://doi.org/10.1016/j.mcat.2021.111645
4. K. Nwosu-Obieogu et al., Melon seed oil epoxidation: kinetics and neuro-fuzzy evaluation. South African J. Chem. Eng. **47**(April 2023), 169–177 (2024). https://doi.org/10.1016/j.sajce.2023.11.010
5. M. Zadshir, B.W. Kim, H. Yin, Bio-Based phase change materials for sustainable development. Materials (Basel) **17**(19), 1–16 (2024). https://doi.org/10.3390/ma17194816
6. A. González-Benjumea et al., High epoxidation yields of vegetable oil hydrolyzates and methyl esters by selected fungal peroxygenases. Front. Bioeng. Biotechnol. **8**(January), 1–12 (2021). https://doi.org/10.3389/fbioe.2020.605854
7. M. Flores et al., Edible oil parameters during deterioration processes. Int. J. Food Sci. **2021** (2021). https://doi.org/10.1155/2021/7105170

8. H. Jin et al., Application of Raman spectroscopy in the rapid detection of waste cooking oil. Food Chem. **362**(December), 2021 (2020). https://doi.org/10.1016/j.foodchem.2021.130191
9. C. Medeiros Vicentini-Polette, P. Rodolfo Ramos, C. Bernardo Gonçalves, A. Lopes De Oliveira, Determination of free fatty acids in crude vegetable oil samples obtained by high-pressure processes. Food Chem. X **12**, 0–7 (2021). https://doi.org/10.1016/j.fochx.2021.100166
10. D. Meyer, S. Hörnlein, D. Rein, G.E. Morlock, Proposed RP–HPTLC–FLD kit for analysis of vitamin A palmitate in edible oils. Anal. Bioanal. Chem. 3791–3801 (2025). https://doi.org/10.1007/s00216-025-05894-0
11. P.T. Wai, P. Jiang, Y. Shen, P. Zhang, Q. Gu, Y. Leng, Catalytic developments in the epoxidation of vegetable oils and the analysis methods of epoxidized products. RSC Adv. **9**(65), 38119–38136 (2019). https://doi.org/10.1039/c9ra05943a
12. P.K. Gamage, M. O'Brien, L. Karunanayake, Epoxidation of some vegetable oils and their hydrolysed products with peroxyformic acid - Optimised to industrial scale. J. Natl. Sci. Found. Sri Lanka **37**(4), 229–240 (2009). https://doi.org/10.4038/jnsfsr.v37i4.1469
13. A.N. Zakwan, R. Siti, F. Nazrah, Comparative study on optimization of factors affecting epoxidation-hydroxylation reaction for the production of waste cooking oil based-polyol. Sci. Res. J. **19**(2), 1–22 (2022). https://doi.org/10.24191/srj.v19i2.13748
14. F. Marriam, A. Irshad, I. Umer, M.A. Asghar, M. Atif, Vegetable oils as bio-based precursors for epoxies. Sustain. Chem. Pharm. **31**(October), 2023 (2022). https://doi.org/10.1016/j.scp.2022.100935
15. J.A.S. Montenegro, A. Ries, I.D.S. Silva, C.B.B. Luna, A.L. Souza, R.M.R. Wellen, Enzymatic and synthetic routes of castor oil epoxidation. Polymers (Basel) **15**(11) (2023). https://doi.org/10.3390/polym15112477
16. A. Gök, H. Uyar, Ö. Demir, Pomegranate seed oil extraction by cold pressing, microwave and ultrasound treatments. Biomass Convers. Biorefinery 0123456789 (2024). https://doi.org/10.1007/s13399-024-05611-4
17. Q.W. Zhang, L.G. Lin, W.C. Ye, Techniques for extraction and isolation of natural products: a comprehensive review. Chinese Med. (United Kingdom) **13**(1), 1–26 (2018). https://doi.org/10.1186/s13020-018-0177-x
18. M.J. González-Fernández, F. Manzano-Agugliaro, A. Zapata-Sierra, E.H. Belarbi, J.L. Guil-Guerrero, Green argan oil extraction from roasted and unroasted seeds by using various polarity solvents allowed by the EU legislation. J. Clean. Prod. **276** (2020). https://doi.org/10.1016/j.jclepro.2020.123081
19. F. Al Juhaimi, N. Uslu, E.E. Babiker, K. Ghafoor, I.A. Mohamed Ahmed, M.M. Özcan, The effect of different solvent types and extraction methods on oil yields and fatty acid composition of safflower seed. J. Oleo Sci. **68**(11), 1099–1104 (2019). https://doi.org/10.5650/jos.ess19131
20. S.E. Onoji, S.E. Iyuke, A.I. Igbafe, Hevea brasiliensis (rubber seed) oil: extraction, characterization, and kinetics of thermo-oxidative degradation using classical chemical methods. Energy Fuels **30**(12), 10555–10567 (2016). https://doi.org/10.1021/acs.energyfuels.6b02267
21. S.H. Mohd-Setapar, L. Nian-Yian, N.S. Mohd-Sharif, Extraction of rubber (hevea brasiliensis) seed oil using soxhlet method. Malaysian J. Fundam. Appl. Sci. **10**(1) (2014). https://doi.org/10.11113/mjfas.v10n1.61
22. P. Raknam, S. Pinsuwan, T. Amnuaikit, Rubber seed cleansing oil formulation and its efficacy of makeup remover. Int. J. Pharm. Sci. Res. **11**(1), 146 (2020). https://doi.org/10.13040/IJPSR.0975-8232.11(1).146-55
23. M.A. Alam et al., Comparative evaluation of physicochemical and antimicrobial properties of rubber seed oil from different regions of Bangladesh. Heliyon **10**(4), e25544 (2024). https://doi.org/10.1016/j.heliyon.2024.e25544
24. H. Suwari, Z. Kotta, Y. Buang, Optimization of soxhlet extraction and physicochemical analysis of crop oil from seed kernel of Feun Kase (Thevetia peruviana). AIP Conf. Proc. **1911** (2017). https://doi.org/10.1063/1.5015998
25. A.S. Reshad, P. Tiwari, V.V. Goud, Extraction of oil from rubber seeds for biodiesel application: optimization of parameters. Fuel **150**, 636–644 (2015). https://doi.org/10.1016/j.fuel.2015.02.058

26. I. Efthymiopoulos et al., Influence of solvent selection and extraction temperature on yield and composition of lipids extracted from spent coffee grounds. Ind. Crops Prod. **119**(April), 49–56 (2018). https://doi.org/10.1016/j.indcrop.2018.04.008
27. M. Mohsen-Nia, H. Amiri, B. Jazi, Dielectric constants of water, methanol, ethanol, butanol and acetone: measurement and computational study. J. Solution Chem. **39**(5), 701–708 (2010). https://doi.org/10.1007/s10953-010-9538-5
28. H. Nawaz, M.A. Shad, N. Rehman, H. Andaleeb, N. Ullah, Effect of solvent polarity on extraction yield and antioxidant properties of phytochemicals from bean (Phaseolus vulgaris) seeds. Brazilian J. Pharm. Sci. **56** (2020). https://doi.org/10.1590/s2175-97902019000417129
29. M.J. Zarrinmehr et al., The effect of solvents polarity and extraction conditions on the microalgal lipids yield, fatty acids profile, and biodiesel properties. Bioresour. Technol. **344**(November), 2022 (2021). https://doi.org/10.1016/j.biortech.2021.126303
30. N. Hambali, M.J. Jalil, I.S. Azmi, A. Shamjuddin, N.H. Alias, Extraction of rubber seed oil as a feedstock for epoxidation process via mixed solvent. Biomass Convers. Biorefinery 0123456789 (2025). https://doi.org/10.1007/s13399-025-06552-2
31. A. Abdolshahi, M.H. Majd, J.S. Rad, M. Taheri, A. Shabani, J.A. Teixeira da Silva, Choice of solvent extraction technique affects fatty acid composition of pistachio (Pistacia vera L.) oil. J. Food Sci. Technol. **52**(4), 2422–2427 (2015). https://doi.org/10.1007/s13197-013-1183-8
32. M. Yang, W. Zhu, H. Cao, Biorefinery methods for extraction of oil and protein from rubber seed (2021)
33. Z. Wang et al., Optimization of oil extraction from rice bran with mixed solvent using response surface methodology. Foods **11**(23), 1–15 (2022). https://doi.org/10.3390/foods11233849
34. C.J. Li, M.R. Xin, Z.L. Sun, Selection of extraction solvents for edible oils from microalgae and improvement of the oxidative stability. J. Biosci. Bioeng. **132**(4), 365–371 (2021). https://doi.org/10.1016/j.jbiosc.2021.06.008
35. Â.C. Salvador, M.M.Q. Simões, A.M.S. Silva, S.A.O. Santos, S.M. Rocha, A.J.D. Silvestre, Vine waste valorisation: Integrated approach for the prospection of bioactive lipophilic phytochemicals. Int. J. Mol. Sci. **20**(17) (2019). https://doi.org/10.3390/ijms20174239
36. Z. Szabo et al., Plant-Based cooking oils (2022)
37. L. Geng, K. Liu, H. Zhang, Lipid oxidation in foods and its implications on proteins. Front. Nutr. **10**(June), 1–12 (2023). https://doi.org/10.3389/fnut.2023.1192199
38. E. Parra-Ortiz et al., Effects of oxidation on the physicochemical properties of polyunsaturated lipid membranes. J. Colloid Interface Sci. **538**, 404–419 (2019). https://doi.org/10.1016/j.jcis.2018.12.007
39. D.K. Chu et al., Physical distancing, face masks, and eye protection to prevent person-to-person transmission of SARS-CoV-2 and COVID-19: a systematic review and meta-analysis. Lancet **395**(10242), 1973–1987 (2020). https://doi.org/10.1016/S0140-6736(20)31142-9
40. I. Čechovičienė et al., Supercritical CO_2 and conventional extraction of bioactive compounds from different cultivars of blackberry (Rubus fruticosus L.) pomace. Plants **13**(20) (2024). https://doi.org/10.3390/plants13202931
41. N.A. Ibrahim, M.A.A. Zaini, Microwave-assisted solvent extraction of castor oil from castor seeds. Chinese J. Chem. Eng. **26**(12), 2516–2522 (2018). https://doi.org/10.1016/j.cjche.2018.07.009
42. E.C. Creencia, J.A.P. Nillama, I.L. Librando, Microwave-assisted extraction and physicochemical evaluation of oil from Hevea brasiliensis seeds. Resources **7**(2) (2018). https://doi.org/10.3390/resources7020028
43. T. Boyapati, S.S. Rana, P. Ghosh, Microwave-assisted extraction of dragon fruit seed oil: fatty acid profile and functional properties. J. Saudi Soc. Agric. Sci. **22**(3), 149–157 (2023). https://doi.org/10.1016/j.jssas.2022.08.001
44. R.C.N. Thilakarathna, L.F. Siow, T.K. Tang, Y.Y. Lee, A review on application of ultrasound and ultrasound assisted technology for seed oil extraction. J. Food Sci. Technol. **60**(4), 1222–1236 (2023). https://doi.org/10.1007/s13197-022-05359-7
45. L. Shen et al., A comprehensive review of ultrasonic assisted extraction (UAE) for bioactive components: principles, advantages, equipment, and combined technologies. Ultrason. Sonochem. **101**(September) (2023). https://doi.org/10.1016/j.ultsonch.2023.106646

46. H. Teng et al., Ultrasonic-assisted extraction of raspberry seed oil and evaluation of its physico-chemical properties, fatty acid compositions and antioxidant activities. PLoS ONE **11**(4), 1–17 (2016). https://doi.org/10.1371/journal.pone.0153457

Chapter 6
Catalytic Epoxidation of Palm Oil and Waste Cooking Oil Using Heterogeneous Catalyst

6.1 Introduction

The shift toward greener and more sustainable chemical processes has become increasingly urgent in recent decades. Among the many approaches under exploration, the use of renewable resources like vegetable oils to produce valuable intermediates such as epoxides has shown great promise [1]. Epoxidized vegetable oils are attractive not only for their renewable origin but also because of their broad applications in polymer synthesis, plasticizers, stabilizers, and coatings [2]. Central to this approach is the epoxidation reaction, which introduces reactive oxirane rings into unsaturated fatty acid chains, offering a versatile platform for further chemical modification. A wide range of vegetable oils, including soybean, sunflower, linseed, castor, and palm oil, have been investigated for epoxidation [3, 4]. Palm oil, in particular, is favored in Southeast Asia due to its high oleic content and year-round availability [5]. However, the fact remains that many of these oils are edible and form part of the global food supply. Using food-grade oils for industrial purposes raises ethical concerns and has driven researchers to explore alternative feedstocks that are both sustainable and non-competitive with food systems.

One such alternative is waste cooking oil (WCO). Generated in large volumes from households and the food industry, WCO is typically discarded despite being rich in unsaturated fatty acids. Its reuse in chemical processes represents a meaningful step toward circular economy practices. However, WCO poses several challenges for epoxidation. High free fatty acid (FFA) content, thermal degradation products, and polymerized triglycerides make WCO a more reactive and less predictable feedstock compared to fresh oil [6]. These factors often lead to lower oxirane oxygen content (OOC) and higher chances of side reactions. Several studies have reported this limitation. For instance, Borugadda and Goud (2014) found that WCO epoxidation using a homogeneous acid system yielded only about 6.2% relative conversion to oxirane (RCO) [7]. Similarly, Silviana et al. [8] observed around 20% RCO when inorganic acids were used as catalysts. These values are significantly lower than

M. J. Jalil et al., *Bio-Based Epoxides*,
SpringerBriefs in Applied Sciences and Technology,
https://doi.org/10.1007/978-981-95-5578-9_6

those achieved with fresh vegetable oils, which can exceed 80% under optimized conditions [9]. To overcome this, blending WCO with a more stable and reactive oil like palm oil has emerged as a promising strategy. The idea is to create a hybrid feedstock that retains the sustainability benefits of WCO while improving its chemical performance through the inclusion of refined palm oil. This not only helps reduce the environmental burden associated with palm monocultures but also enhances oxirane formation by introducing more structurally consistent triglycerides.

At the same time, the choice of catalyst plays a crucial role in determining how efficiently the epoxidation proceeds. While traditional mineral acids like sulfuric acid are effective, they bring problems of corrosion, difficult separation, and environmental impact [10, 11]. Solid acid catalysts like Amberlite IR-120H, a sulfonated polystyrene-divinylbenzene resin, offer a more sustainable and practical alternative. Amberlite catalyzes the in situ formation of performic or peracetic acid from hydrogen peroxide and organic acids, which then epoxidize the double bonds under relatively mild conditions. The use of Amberlite IR-120H in vegetable oil epoxidation is well supported in the literature. Nuruddin et al. [12] reported its effectiveness in catalyzing the epoxidation of oleic acid derived from WCO, achieving notable OOC values and minimal ring degradation. Azmi et al. [13] and Jalil et al. [14] demonstrated similar success with palm-based systems, confirming good selectivity and retention of catalytic activity across multiple cycles. These studies highlight Amberlite's suitability for greener, more robust epoxidation protocols. Yet, while there is growing interest in solid acid-catalyzed epoxidation, few studies have looked at hybrid feedstocks, particularly palm oil mixed with WCO in combination with Amberlite. This gap is important because hybrid systems offer a practical compromise between sustainability and reactivity, but their performance under heterogeneous catalysis remains largely unexplored.

In this study, the catalytic epoxidation of a hybrid blend consisting of palm oil and waste cooking oil (WCO) was carried out using Amberlite IR-120H as a heterogeneous catalyst. The investigation focused on the influence of key reaction parameters on oxirane oxygen content (OOC) and relative conversion to oxirane (RCO). Specifically, the effects of stirring speed (150, 250, and 350 rpm) and formic acid to oil molar ratio (1:1, 1.5:1, and 2:1) were systematically studied under controlled conditions. These parameters were selected due to their critical roles in mass transfer efficiency and peracid formation, which directly impact the selectivity and extent of epoxidation. The outcomes are expected to provide insights into optimizing hybrid feedstock epoxidation systems using solid acid catalysts, while contributing to the sustainable valorization of waste oils for industrial applications.

6.2 Methodology

6.2.1 *Materials*

The waste cooking oil (WCO) used in this study was collected from a local restaurant in Pasir Gudang, Malaysia, following standard filtration to remove food residues and particulates. The palm oil was purchased from a local supermarket under the commercial brand Alif (manufactured in Malaysia) and used without further modification. The reagents employed in the epoxidation process were of analytical grade. Hydrogen peroxide (30% w/w), formic acid ($\geq$95%), hydrobromic acid (48% in acetic acid), crystal violet indicator, and glacial acetic acid ($\geq$99.7%) were all obtained from Sigma-Aldrich (USA). The Amberlite IR-120H ion-exchange resin, used as the heterogeneous catalyst, was also supplied by Sigma-Aldrich. All chemicals were used as received without further purification unless otherwise stated. Deionized water was used in all solution preparations and cleaning procedures.

6.2.2 *Experimental Setup and Procedure*

The epoxidation reactions were carried out in a 500 mL round-bottom flask immersed in a thermostatically controlled water bath to maintain a stable reaction temperature. A mechanical overhead stirrer was used to ensure consistent mixing of the reactants and catalyst. A total of 100 g of hybrid oil blend prepared by mixing palm oil and waste cooking oil (WCO) in a 1:1 weight ratio was charged into the flask. The mixture was gradually heated to 80 °C. Formic acid was added directly into the reactor at molar ratios of 1:1, 1.5:1, or 2:1 relative to the oil. Immediately following this, hydrogen peroxide (30% w/w) was introduced at a fixed 1:1 molar ratio to oil. Finally, 0.5 g of Amberlite IR-120H resin was added to the mixture to initiate the in situ generation of performic acid.

The reaction was allowed to proceed for 180 min under continuous stirring. Three different stirring speeds (150, 250, and 350 rpm) were investigated to assess the effect of mixing intensity on oxirane formation. At 30-min intervals, approximately 5 mL of the reaction mixture was withdrawn using a pipette and immediately cooled in an ice bath to quench the reaction. The samples were filtered to separate the catalyst and stored in amber bottles prior to analysis. The oxirane oxygen content (OOC) of each sample was determined using the AOCS Official Method Cd 9–57, a titrimetric procedure based on the reaction of oxirane rings with hydrobromic acid in glacial acetic acid, using crystal violet as the indicator. All experiments were performed in triplicate, and average values were reported to ensure reproducibility.

6.2.3 Oxirane Oxygen Content (OOC) and Relative Conversion (RCO) Determination

The determination of oxirane oxygen content (OOC) and the relative conversion to oxirane (RCO) was carried out to evaluate the efficiency of the epoxidation process. The RCO value was derived by comparing the experimental oxirane content with its theoretical counterpart, as described in Eq. 6.1.

$$\text{RCO} = \frac{\text{OOC}_{\text{experiment}}}{\text{OOC}_{\text{theoretical}}} \times 100 \tag{6.1}$$

The theoretical OOC was calculated based on the initial iodine value (X_o), the molar mass of iodine (A_i), and the atomic mass of oxygen (A_o), using the expression shown in Eq. (6.2). This estimation assumes complete conversion of double bonds to oxirane rings and accounts for the stoichiometric relationship between iodine and oxygen atoms.

$$\text{OOC}_{\text{the}} = \left\{\left(\frac{X_o}{A_i}\right) / \left[100 + \left(\frac{X_o}{2A_i}\right)\right]\right\} \times A_o \times 100 \tag{6.2}$$

The experimental OOC, on the other hand, was obtained through titration with hydrobromic acid (HBr), where the volume difference between the titrated sample (V) and the blank (B) reflects the oxirane content. The calculation, as presented in Eq. (6.3), incorporates a constant factor (1.6), the normality of HBr (N), and the mass of the sample (W) to determine the actual oxirane oxygen content. These parameters collectively enabled the quantitative assessment of the epoxidation reaction's effectiveness.

$$\text{OOC}_{\text{Exp}} = 1.6 \times N \times \frac{(V - B)}{W} \tag{6.3}$$

6.2.4 Fourier Transform Infrared (FTIR) Spectroscopy Analysis

Fourier Transform Infrared (FTIR) spectra were recorded in transmittance mode using a PerkinElmer Spectrum instrument equipped with a deuterated triglycine sulfate (DTGS) detector and a KBr beam splitter. The instrument was operated over the wavenumber range of 4000–650 cm^{-1} with a spectral resolution of 4 cm^{-1}, and each spectrum was obtained by averaging 32 scans to enhance the signal-to-noise ratio. Background spectra were collected before each measurement to ensure baseline accuracy. Liquid samples were prepared as thin films by placing a drop of the sample between two polished potassium bromide (KBr) windows. All spectra were baseline-corrected and analyzed to identify characteristic functional groups.

6.3 Results and Discussions

6.3.1 Effect of Stirring Speed Toward Epoxidation Process

The epoxidation of the hybrid palm/waste-oil showed the typical trend of rising oxirane content to a peak followed by a decline at longer reaction times. Notably, the stirring speed had a strong influence on both the rate of oxirane formation and the timing of the peak conversion. At the lowest agitation (150 rpm), oxirane oxygen content increased slowly and reached its maximum later than in the faster-stirred runs. In contrast, higher stirring speeds (250 and 350 rpm) accelerated the epoxidation kinetics where oxirane conversion rose more rapidly, and the peak oxirane content was attained in a shorter time. The 350 rpm run exhibited the fastest initial formation of epoxide, with the earliest peak conversion, indicating that intensive agitation enhances the in situ generation and diffusion of peracid (formed from H_2O_2 and carboxylic acid) into the oil phase. This is consistent with the notion that stirring speed improves mass transfer: vigorous mixing ensures efficient contact between the reactants (oil, acid, H_2O_2) and the Amberlite IR-120H catalyst, promoting faster peracid formation and delivery to the double bonds. Consequently, higher stirring yielded a greater oxirane conversion in the early stages compared to 150 rpm (Fig. 6.1).

Interestingly, although all systems displayed an initial rise in RCO, a decline was observed after reaching the peak. This decline, more pronounced at lower rpm, may be due to ring-opening side reactions where excess hydrogen peroxide reacts with the oxirane ring, forming diols or glycol derivatives. These secondary reactions tend to dominate under prolonged reaction times or insufficient mixing, emphasizing the importance of optimizing both duration and agitation speed.

The findings are consistent with trends reported in the literature for solid acid–catalyzed epoxidation systems, and they highlight the advantages of faster kinetics and efficient mass transfer. Many studies on vegetable oil epoxidation emphasize that vigorous agitation is needed to overcome mass transfer limitations between the immiscible phases. The epoxidation proceeds via in situ generated peracids (performic or peracetic acid) reacting with the oil's $C = C$ bonds; this occurs at the interface of an aqueous phase (where peracid forms) and the oil phase [15]. Insufficient stirring leads to a diffusion barrier that slows the delivery of oxidants to the oil, limiting the reaction rate.

Literature on solid acid catalysts in epoxidation also underscores the balance between reaction rate and product stability. Rahim et al. [16] observed that overly high stirring speeds, while accelerating the initial epoxide formation, can also accelerate side reactions that degrade the epoxide product. In their study on epoxidation of palm stearin with a heterogeneous catalyst, very fast agitation (e.g., 500 rpm in

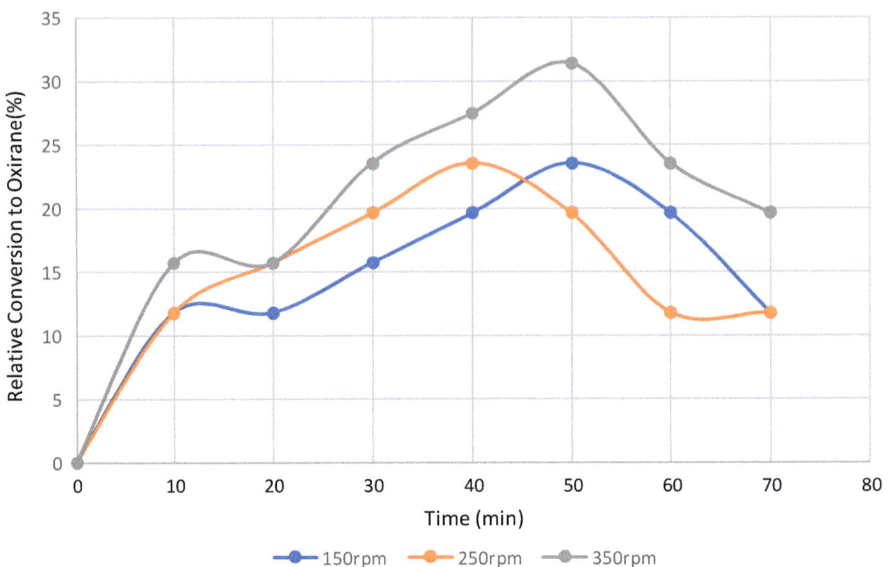

Fig. 6.1 Relative conversion to oxirane versus time for effect of stirring speed on epoxidation process

their system) led to a quicker attainment of peak oxirane but then a faster decline due to shear-induced or enhanced ring-opening reaction. They concluded that a moderate stirring speed provided the best compromise, allowing high conversion without excessive product degradation. This aligns with the result that 350 rpm is an optimal, moderate-high speed which is vigorous enough to drive fast conversion, but not so extreme as to cause uncontrolled turbulence or premature epoxide breakdown.

It is worth noting that the hybrid use of palm oil and waste cooking oil could have introduced varying degrees of unsaturation and oxidation stability, which also affects RCO performance. Waste cooking oil, typically degraded through repeated heating, might carry peroxides or free fatty acids that either accelerate or inhibit the epoxidation process depending on the reaction environment. Overall, 350 rpm appears to be the optimum stirring speed for maximizing oxirane yield under the studied conditions. However, excessive agitation beyond this point may cause emulsification issues or mechanical stress on the resin, as suggested by previous studies.

6.3.2 Effect of Formic Acid to Hybrid Oil Molar Ratio Toward Epoxidation Process

The effect of formic acid to oleic acid (FA: OA) molar ratio on the relative conversion to oxirane (RCO) was evaluated using a hybrid feedstock composed of palm oil and waste cooking oil. As illustrated in Fig. 6.2, the conversion trends over time were found to be significantly influenced by the FA: OA molar ratio. Among the three ratios tested, the molar ratio of 1:1 exhibited the highest RCO, peaking at approximately 39% at 35 min. This result suggests that a balanced molar ratio of formic acid to double bonds is crucial in promoting efficient in situ formation of performic acid, which subsequently reacts with the carbon–carbon double bonds to form epoxide groups.

Mechanistically, this behavior reflects the dual role of formic acid in the in situ epoxidation via performic acid. Formic acid is necessary to generate the performic acid oxidizing agent, but an optimal amount exists. Insufficient formic acid (below 1:1) limits the rate of peracid formation, slowing epoxidation and yielding less oxirane. Excess formic acid, on the other hand, creates a highly acidic reaction that accelerates oxirane ring-opening side reactions [17, 18]. When hydrogen peroxide and formic acid react to form performic acid, each mole of peracid generated also produces a mole of water which increases the water content in the mixture. In a strongly acidic, water-rich environment, the freshly formed epoxide rings are prone to acid-catalyzed hydrolysis (ring-opening), forming diols or glycol by-products rather than stable epoxides. Formic acid can also directly participate in nucleophilic ring-opening especially at high concentrations of acid/peracid [19]. These undesired side reactions become kinetically significant at higher FA:oil ratios, effectively

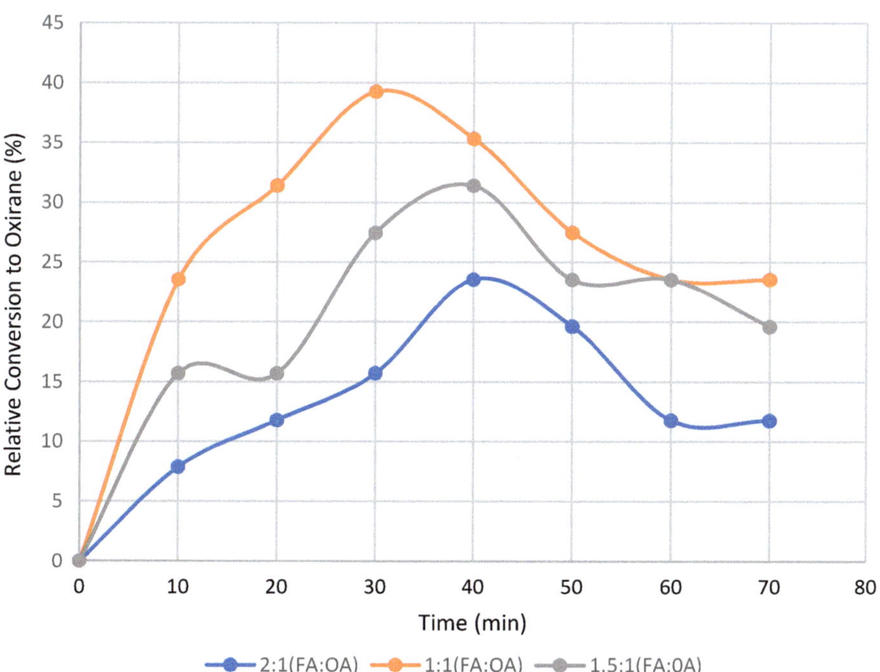

Fig. 6.2 Relative conversion to oxirane versus time for effect of formic acid to hybrid oil molar ratio toward epoxidation process

consuming the oxirane rings almost as quickly as they form. The net effect is that beyond a certain acid threshold, additional formic acid speeds up epoxide formation and destruction simultaneously, leading to a lower observable oxirane peak and faster decline.

It is important to note that the use of a solid acid catalyst, Amberlite IR-120H, helps moderate the reactivity of the system to some extent. Amberlite IR-120H is a sulfonic acid ion-exchange resin that provides protons heterogeneously; it absorbs and holds the performic acid within its porous structure. This localization can reduce oxirane degradation by limiting the bulk concentration of free formic acid/peracid and by temporarily sequestering the epoxidation intermediate within the resin matrix. In fact, Rahman et al. [17] reported that Amberlite's ability to bind performic acid slows down oxirane ring opening compared to a homogeneous acid system. However, even with Amberlite, an overload of formic acid will overwhelm the benefits of the heterogeneous catalyst. Figure 6.2 indicates that at 1.5:1 and especially 2:1 FA:oil, the intrinsic acidity in the system is too high, leading to rapid epoxy ring cleavage despite Amberlite's presence. The resin cannot fully protect epoxides when formic acid is in large excess, because the rate of ring-opening (catalyzed by surplus acid and water) begins to outpace the rate of epoxide formation. Thus, beyond the optimal 1:1 ratio, additional formic acid undermines oxirane yield by promoting these side reactions.

This trend is consistent with findings in literature for epoxidation of vegetable oils with in situ generated peracids. Epoxidation is maximized at a moderate acid level that achieves efficient peracid generation but avoids overly acidic conditions. For instance, a study on kapok seed oil shows that a formic acid to double-bond ratio of 0.5 produced the highest oxirane oxygen content, while increasing the ratio to 1.0 showed no further gain; instead, epoxide selectivity dropped due to ring-opening reactions [19]. In that system, more than half of the double bonds converted were ultimately diverted to side-products (diols, hydroxy-esters, etc.) rather than epoxides at high acid levels. These side reactions (epoxy ring opening by performic acid, carboxylic acid, water, and even residual H_2O_2) have been widely documented to limit epoxide yields in acid-catalyzed epoxidations. Even strong homogeneous acids, like sulfuric acid, while accelerating the epoxidation rate, are known to exacerbate oxirane cleavage, yielding glycols and other by-products that lower the final oxirane content. The advantage of a solid acid like Amberlite is mainly in mitigating corrosion and somewhat enhancing selectivity, but the fundamental kinetic trade-off remains: an overly acidic/reactive medium will convert epoxides to ring-opened products as soon as they form.

Considering these insights, the superior performance of the 1:1 FA: oil ratio in Fig. 6.2 can be attributed to it striking the optimal balance between reactivity and selectivity. At 1:1, the formic acid is sufficient to rapidly produce performic acid and drive the epoxidation of the palm oil/WCO hybrid feedstock, yet the acid level is low enough to keep hydrolysis and ring-opening side reactions to a minimum during the critical epoxidation period. The result is a high oxirane yield (39%) that is attained quickly and maintained for a longer duration, indicative of high selectivity toward epoxide formation.

In contrast, higher FA: oil ratios (1.5:1 and 2:1) tilt the balance too far where they generate performic acid even faster (giving an initial burst of epoxidation) but also introduce excess acidity and water that accelerate the degradation of the very oxirane rings being formed. The net oxirane content peaks lower and then falls as epoxides are consumed by side reactions. Thus, the 1:1 ratio emerges as optimal: it maximizes the in situ peracid formation rate without crossing the threshold where oxirane-opening reactions dominate.

This outcome underscores a key principle for high-yield epoxidation of unsaturated oils where more oxidizing acid is not always better, as the reaction must be carefully tuned to favor epoxide formation over ring opening. By maintaining a balanced FA:oil ratio, the Amberlite-catalyzed epoxidation achieves superior reactivity-selectivity synergy, in agreement with other high-efficiency epoxidation processes reported in the literature. The finding that a 1:1 FA:oil ratio gives the highest oxirane yield highlights the importance of avoiding excess formic acid, since beyond the optimal point each additional increment of acid only serves to accelerate epoxide consumption rather than formation.

6.3.3 Analysis of FTIR

The FTIR spectra of the raw hybrid oil (a mixture of palm oil and waste cooking oil) and the corresponding epoxidized product are compared in Fig. 6.3. The hybrid oil spectrum exhibits characteristic absorption bands of triglycerides, including strong C–H stretching vibrations of aliphatic chains at 2850–2925 cm^{-1}, and a sharp ester carbonyl (C = O) stretching band at ~1740 cm^{-1}, consistent with reported vegetable oil spectra. In addition, a weak band at ~3006 cm^{-1}, attributed to = C–H stretching of cis-olefinic bonds, and a peak at ~1650 cm^{-1}, corresponding to C = C stretching, confirm the presence of unsaturation in the hybrid feedstock.

After epoxidation, significant spectral changes are observed. The disappearance of the = C–H band at ~3006 cm^{-1} and the substantial reduction of the C = C stretching signal at ~1650 cm^{-1} indicate that the double bonds were successfully consumed during the reaction. Simultaneously, new absorptions appear in the range of 820–850 cm^{-1}, assigned to the asymmetric stretching of the oxirane ring, and at ~1240 cm^{-1}, associated with C–O stretching of the epoxide functionality. These bands are widely recognized as the key diagnostic features of oxirane formation in vegetable oil epoxidation.

The retention of the ester carbonyl peak at 1740 cm^{-1} in both spectra suggests that the triglyceride backbone remains intact, with no evidence of ester cleavage. Furthermore, the aliphatic C–H stretching bands remain prominent, demonstrating that the long-chain fatty acid structures are preserved post-epoxidation. These findings align with previous studies on the epoxidation of vegetable oil, which similarly reported the disappearance of unsaturation bands and the emergence of characteristic oxirane absorptions [20–22]. However, the current work extends the understanding by employing a hybrid feedstock system, combining palm oil with waste cooking oil.

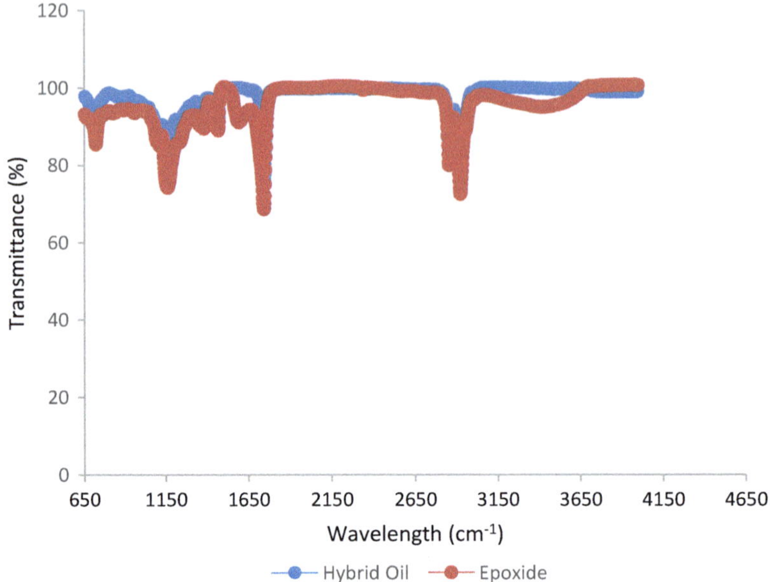

Fig. 6.3 Fourier transform infrared spectroscopy (FTIR) for hybrid oil and epoxide

This blend introduces a broader distribution of fatty acids and unsaturation levels, which can influence epoxidation kinetics and oxirane yield. The successful detection of oxirane bands in this hybrid matrix highlights the feasibility of utilizing mixed oil streams for high-value intermediate production. Overall, the FTIR analysis confirms the effective conversion of double bonds in the hybrid oil to oxirane groups, validating the success of the epoxidation process. The clear spectral evidence of $C = C$ bond depletion and oxirane ring formation provides a robust foundation for subsequent kinetic analysis and process optimization. These results not only demonstrate consistency with established literature but also emphasize the novelty of hybrid oil valorization in epoxide synthesis, supporting the potential for industrial application.

6.4 Conclusion

This study established that the epoxidation of hybrid oil derived from palm oil and waste cooking oil can be efficiently achieved using Amberlite IR-120H in an in situ performic acid system, with process parameters playing a decisive role in maximizing oxirane yield and selectivity. Optimal conditions were identified at a stirring speed of 350 rpm and a formic acid-to-oil molar ratio of 1:1, which delivered the highest oxirane oxygen content while minimizing ring-opening side reactions. FTIR analysis confirmed the successful conversion of unsaturated bonds into oxirane groups, validating the efficacy of the optimized process. These findings highlight the novelty of hybrid oil utilization, demonstrating that strategic blending of waste and refined oils can overcome the limitations of waste cooking oil alone and provide a sustainable route to high-value bio-based epoxides. Future investigations should extend the hybrid oil strategy to other waste-derived and non-edible feedstocks, coupled with detailed kinetic modeling and catalyst reusability assessments, to strengthen mechanistic understanding and advance process scalability. Such efforts will accelerate the transition of epoxidized hybrid oils from laboratory studies to industrial applications, supporting the broader adoption of renewable precursors in polymer and materials industries.

References

1. Y. Zeng, Z. Shang, Z. Zheng, N. Shi, B. Yang, S. Han, J. Yan, A review of chemical modification of vegetable oils and their applications. Lubricants **12** (2024). https://doi.org/10.3390/lubricants12050180
2. J. Thomas, R. Patil, Enabling green manufacture of polymer products via vegetable oil epoxides. Ind. Eng. Chem. Res. **62**, 1725–1735 (2023). https://doi.org/10.1021/acs.iecr.2c03867
3. C.K. Patil, D.W. Jung, H.D. Jirimali, J.H. Baik, V.V. Gite, S.C. Hong, Nonedible vegetable oil-based polyols in anticorrosive and antimicrobial polyurethane coatings. Polymers (Basel). **13**, 1–28 (2021). https://doi.org/10.3390/polym13183149
4. Y. Ma, Y. Fang, Y. Hu, Q. Li, Q. Huang, Q. Shang, M. Zhang, S. Li, P. Jia, Y. Zhou, Recent advances in vegetable oil based fine chemicals and polymers. Green Mater. 1–22 (2024). https://doi.org/10.1680/jgrma.24.00083
5. W.F. Bohórquez, A. Orjuela, P.C.N. Rincón, J.G. Cadavid, J.A. García-Nunez, Experimental optimization during epoxidation of a high-oleic palm oil using a simplex algorithm. Ind. Crops Prod. **187**, 1–8 (2022). https://doi.org/10.1016/j.indcrop.2022.115321
6. M. Zulqarnain Ayoub, M.H.M. Yusoff, M.H. Nazir, I. Zahid, M. Ameen, F. Sher, D. Floresyona, E. Budi Nursanto, A comprehensive review on oil extraction and biodiesel production technologies. Sustainability **13**, 1–28 (2021). https://doi.org/10.3390/su13020788
7. O. Study, V.B. Borugadda, V.V Goud, Synthesis of waste cooking oil epoxide as a bio-lubricant base stock: synthesis of waste cooking oil epoxide as a bio-lubricant base stock: characterization and optimization study (2014). https://doi.org/10.1166/jbeb.2014.1077
8. S. Silviana, D.D. Anggoro, A.C. Kumoro, Waste cooking oil utilisation as bio-plasticiser through epoxidation using inorganic acids as homogeneous. Catalysts **56**, 1861–1866 (2017). https://doi.org/10.3303/CET1756311
9. S. Meadows, M. Hosur, Y. Celikbag, S. Jeelani, Comparative analysis on the epoxidation of soybean oil using formic and acetic acids **26**, 289–298 (2018)
10. Y. Meng, F. Taddeo, A.F. Aguilera, X. Cai, V. Russo, P. Tolvanen, S. Leveneur, The lord of the chemical rings: catalytic synthesis of important industrial epoxide compounds. Catalysts **11** (2021). https://doi.org/10.3390/catal11070765
11. S.K. Tulashie, E.M. Alale, P.Q. Agudah, C.A. Osei, C.A. Munumkum, B.K. Gah, E.B. Baidoo, A review on the production of biodiesel from waste cooking oil: a circular economy approach. Biofuels 1–21 (2024). https://doi.org/10.1080/17597269.2024.2384277
12. D. Nuruddin, A. Raofuddin, I. Suhada, A. Mohd, J. Jalil, Catalytic epoxidation of oleic acid derived from waste cooking oil by in situ peracids. J. Polym. Environ. (2023). https://doi.org/10.1007/s10924-023-02978-9
13. I.S. Azmi, M.J. Jalil, A. Hadi, Epoxidation of unsaturated fatty acid–based palm oil via peracid mechanism as an intermediate product. Biomass Convers. Biorefinery. **14**, 7847–7855 (2024). https://doi.org/10.1007/s13399-022-02862-x

14. M. Jumain, J. Intan, S. Azmi, A. Farhan, M. Yamin, M. Zarqani, Y. Abdul, Catalytic epoxidation of oleic acid and subsequent ring—opening by in situ hydrolysis for production dihydroxystearic acid. J. Polym. Environ. (2022). https://doi.org/10.1007/s10924-022-02736-3
15. J. Chen, M. De Liedekerke Beaufort, L. Gyurik, J. Dorresteijn, M. Otte, R.J.M. Klein Gebbink, Highly efficient epoxidation of vegetable oils catalyzed by a manganese complex with hydrogen peroxide and acetic acid. Green Chem. **21,** 2436–2447 (2019). https://doi.org/10.1039/c8gc03857k
16. N.H. Rahim, M.J. Jalil, N.M. Mubarak, I.S. Azmi, G. Anbuchezhiyan, Catalytic epoxidation of unsaturated fatty acids in palm stearin via in situ peracetic acids mechanism. Sci. Rep. **15**, 1–12 (2025). https://doi.org/10.1038/s41598-025-89399-x
17. M.A. Rahman, N.M. Mubarak, I.S. Azmi, M.J. Jalil, Sustainable approach for catalytic green epoxidation of oleic acid with applied ion exchange resin. Sci. Rep. **13**, 1–8 (2023). https://doi.org/10.1038/s41598-023-42879-4
18. E.R. Gunawan, D. Suhendra, P. Arimanda, D. Asnawati, Murniati, Epoxidation of Terminalia catappa L. Seed oil: Optimization reaction. South African J. Chem. Eng. **43**, 128–134 (2023). https://doi.org/10.1016/j.sajce.2022.10.011
19. I.D.G.A. Putrawan, A. Azharuddin, J. Jumrawati, Preparing epoxidized vegetable oil from waste generated by the kapok fiber industry and assessing its thermal stabilization effect as a co-stabilizer for polyvinyl chloride. Heliyon. **9**, e19624 (2023). https://doi.org/10.1016/j.heliyon.2023.e19624
20. M. Saad, B. Siyo, H. Alrakkad, Preparation and characterization of biodiesel from waste cooking oils using heterogeneous Catalyst(Cat.TS-7) based on natural zeolite, Heliyon. 9, e15836 (2023). https://doi.org/10.1016/j.heliyon.2023.e15836
21. I. Dominguez-Candela, A. Lerma-Canto, S.C. Cardona, J. Lora, V. Fombuena, Physicochemical characterization of novel epoxidized vegetable oil from chia seed oil. Materials (Basel). **15**, 1–19 (2022). https://doi.org/10.3390/ma15093250
22. P.D. Jadhav, A.V. Patwardhan, R.D. Kulkarni, Kinetic study of in situ epoxidation of mustard oil. Mol. Catal. **511**, 111748 (2021). https://doi.org/10.1016/j.mcat.2021.111748

Chapter 7
Optimization and Kinetic Modeling of Catalytic Epoxidation of Palm Stearin Using an In Situ Peracid Mechanism with Sulfuric Acid Catalyst

7.1 Introduction

The global industry is progressively shifting from dependence on non-renewable, petroleum-derived resources to more sustainable, bio-based alternatives. Vegetable oils have the intention to be of considerable interest as a crucial raw material for the oleochemical industry owing to their availability, cost-effectiveness, and potential for sustainable applications [1–3]. Epoxidation is a conversion that incorporates an oxygen atom into a carbon–carbon double bond, thereby converting unsaturated fatty acids into epoxides, which are cyclic ethers characterized by their elevated reactivity. Epoxidized vegetable oils are extensively utilized in the manufacturing of industrial chemicals and materials, including plasticizers for PVC, lubricants, and precursors for polyols, glycols, and polymers such as polyesters and polyurethanes [4–6]. The in situ peracid mechanism for epoxidation, usually conducted in a two-phase system comprising an organic phase and an aqueous phase, offers a promising method. This method provides a sustainable and efficient approach to epoxidizing palm stearin, potentially expanding its applications across various industries. The first step is when an oxygen carrier, like acetic acid or formic acid, reacts with hydrogen peroxide in the presence of an acid catalyst, like sulfuric acid. This makes the corresponding peroxyacid. In the second step, the peroxyacid migrates to the organic phase, where it electrophilically attacks the carbon–carbon double bonds of the unsaturated fatty acids in palm stearin triglycerides, forming the desired oxirane rings [4, 5, 7]. This process, a variant of the Prilezhaev reaction, is commonly catalyzed by sulfuric acid, which efficiently accelerates the rate-determining step, the formation of the peracid itself.

Extensive research has been conducted on the epoxidation of various vegetable oils, including soybean oil, neem oil, palm kernel oil, and crude palm oil. Much of this research has focused on optimizing critical process parameters such as temperature, catalyst concentration, and reactant molar ratios to maximize epoxide yield and efficiency [8, 9]. Statistical methods, particularly the Taguchi technique, have been

M. J. Jalil et al., *Bio-Based Epoxides*,
SpringerBriefs in Applied Sciences and Technology,
https://doi.org/10.1007/978-981-95-5578-9_7

widely employed to streamline the optimization process by reducing the number of experimental runs while still identifying the most influential factors. For instance, studies on neem oil have shown that reaction time is a key factor influencing epoxide yield [10]. Kinetic modeling has also become an essential tool in understanding the underlying reaction mechanisms, predicting process behavior, and facilitating the scale-up of production. However, many existing models are simplified and fail to account for the complexities of two-phase systems, including mass transfer limitations and side reactions like oxirane ring cleavage [11–13]. Despite the substantial body of literature, a gap remains in studies that combine robust statistical optimization approaches, such as Taguchi and ANOVA, with the development of accurate kinetic models specifically for the sulfuric acid-catalyzed epoxidation of palm stearin. This research aims to bridge that gap by precisely identifying the optimal process parameters and providing a predictive mathematical model that aligns closely with experimental data. By addressing this gap, the study offers a powerful tool for process control and provides a deeper understanding of the industrial application of sustainable epoxidation processes.

The main objectives of this study are threefold: (1) to apply the Taguchi experimental design methodology to determine the optimal levels of key process parameters—reaction temperature, stirrer speed, and catalyst loading—to maximize the relative conversion to oxirane (RCO) during the in situ epoxidation of palm stearin; (2) to use Analysis of Variance (ANOVA) to quantify the percentage contribution and statistical significance of each process parameter, identifying the most influential factors governing the reaction; and (3) to develop a mathematical kinetic model using the 4th-order Runge–Kutta technique that accurately simulates the reaction profile, demonstrating a strong correlation with experimental data, and providing a robust predictive tool for the process.

7.2 Material and Methods

7.2.1 Materials

Palm stearin (PS) was sourced from Delima Oil Sdn Bhd, located in Pasir Gudang. Acetic acid (AA) and hydrogen peroxide (HP) were obtained from Qrec Sdn. Bhd., as detailed in Table 7.1. The zeolite ZSM-5 was provided by Fisher Scientific (M) Sdn Bhd. To optimize the epoxidation process of palm stearin, the Taguchi Method was employed using Minitab 18, which facilitated the analysis of the effects of various variables. Additionally, MATLAB was utilized to develop kinetic models that describe the underlying reaction mechanisms.

Table 7.1 List of chemicals used for production of epoxidized palm stearin

Materials	Range/amount	Purity/molarity	Supplier
Palm stearin	100 g		Delima Oil Sdn BhD
Acetic acid molar ratio	0.5, 1.0, 1.5 2.0	23.6 M	QReC Sdn. Bhd
Hydrogen peroxide molar ratio	0.5, 1.0, 1.5 2.0	50%	QReC Sdn. Bhd
Sulfuric acid	5 g		QReC Sdn. Bhd

7.2.2 Experimental for Epoxide Palm Stearin

The epoxidation of palm stearin (PS) to produce epoxidized palm stearin (EPS) was carried out in a 500-mL beaker under a fume hood, using a hot plate for heating. Initially, 100 g of liquid palm stearin was carefully weighed and combined with acetic acid in predetermined molar ratios of 0.5, 1.0, 1.5, and 2.0. The mixture was then transferred into the beaker, which was fitted with a magnetic stirrer and a thermometer. The beaker was placed in a water bath to maintain a stable temperature throughout the reaction. The palm stearin was heated to the desired temperatures of 65 °C, 75 °C, and 85 °C, while being stirred at speeds of 100, 150, and 250 rpm. Hydrogen peroxide was gradually added to the acetic acid-palm stearin mixture in the selected molar ratios of 0.5, 1.0, 1.5, and 2.0. Afterward, varying amounts of homogeneous catalysts (0.2, 0.3, and 0.4 g) were introduced into the mixture. These parameter ranges were chosen based on a thorough review of the literature, which provided established values for key reaction conditions such as temperature, molar ratios, catalyst concentration, and hydrogen peroxide molar ratio [14, 15]. The reaction was allowed to proceed for 70 min, after which a 3 g sample was extracted using a syringe for oxirane concentration analysis, following the AOCS Official Method Cd-957.

7.2.3 Optimization by Taguchi Method

The Taguchi design methodology will be employed to optimize the epoxidation process while minimizing the number of experimental runs. The key parameters under investigation include (A) reaction temperature, (B) catalyst loading (in grams), and (C) stirrer speed (rpm). A comprehensive experimental design, as outlined in Table 7.2, incorporates five parameters at three distinct levels, facilitating an efficient optimization process.

The (RCO) from each experiment was assessed using the signal-to-noise (S/N) ratio and Analysis of Variance (ANOVA) to determine the optimal and most reliable parameter settings for the epoxidation of palm stearin. These approaches were used to identify the conditions that achieve the best results with minimal variability. The experimental design created by Taguchi is presented in Table 7.3.

Table 7.2 The three parameters at three levels for optimizing the epoxidation process

Parameter	Level		
	1	2	3
(A) Reaction temperature (°C)	65	75	85
(B) Catalyst loading (g)	0.2	0.3	0.4
(C) Stirrer speed (rpm)	100	150	250

Table 7.3 The experimental design developed by Taguchi method

No	(A) Reaction temperature (°C)	(B) Catalyst loading (g)	(C) Stirrer speed (rpm)
1	65	0.2	100
2	65	0.3	150
3	65	0.4	250
4	75	0.2	150
5	75	0.3	250
6	75	0.4	150
7	85	0.2	250
8	85	0.3	100
9	85	0.4	150

7.2.4 Relative Conversion Oxirane (RCO)

The experiments were conducted to assess the relative conversion to oxirane (RCO) by comparing the theoretical oxirane oxygen content (OOC) with the experimentally obtained OOC through direct titration using hydrobromic acid (HBr). This technique serves as an effective measure of the unsaturation level in palm stearin. The RCO is then calculated based on the OOC values, which are determined both theoretically and experimentally, as illustrated in Eqs. 7.1–7.3.

$$\mathrm{RCO} = \frac{\mathrm{OOC}_{\mathrm{experimental}}}{\mathrm{OOC}_{\mathrm{theoretical}}} \times 100 \tag{7.1}$$

$$\mathrm{OOC}_{\mathrm{theoretical}} = \left\{\left(\frac{X_0}{A_i}\right) / \left[100 + \left(\frac{X_0}{2A_i}\right)(A_o)\right]\right\} \times A_o \times 100 \tag{7.2}$$

$$\mathrm{OOC}_{\mathrm{experimental}} = 1.6 \times N \times \frac{(V - B)}{W} \tag{7.3}$$

Here, X_0 starting iodine value, A_i molar mass of iodine, A_o molar mass of oxygen, N normality of HBr, V volume of the HBr solution used for the blank in milliliters (mL), V volume of HBr solution used for titration, and W represents the weight of the sample.

7.2.5 Fourier Transform Infrared (FTIR)

FTIR is a method for identifying the functional groups in organic molecules by analyzing their vibrational frequencies in the infrared part of the electromagnetic spectrum [16]. This technique allows us to detect the unique vibration frequencies of different functional groups, making them easier to identify [17]. When infrared radiation passes through an organic compound, it causes the functional groups to vibrate at specific frequencies. An FTIR spectrometer (Spectrum One, Perkin Elmer, USA) was used to identify the functional groups in the samples. This instrument measures infrared radiation from 400 to 4000 cm^{-1}, capturing the basic vibrational modes of the molecules.

7.2.6 Kinetic Modeling of the In Situ Epoxidation Palm Stearin

A model based on assumptions for epoxidation kinetics is framed based on the reaction occurring in a single phase, constant phase volume during the reaction, and no heat transfer considered. The kinetics for a multi-stage reaction system, which includes acetic acid (AA), hydrogen peroxide (HP), palm stearin (PS), epoxidized palm stearin (EPS), and two main reactions in the system: the in situ formation of performic acid (PA) (Eq. 7.4) and the subsequent formation of EPS (Eq. 7.5) are taken.

$$\mathrm{AA} + \mathrm{HP} \underset{k_{12}}{\overset{\rightarrow^{k_{11}}}{\leftarrow}} \mathrm{PA} + \mathrm{Water} \tag{7.4}$$

$$\mathrm{PA} + \mathrm{PS} \rightarrow^{k_2} \mathrm{EPS} + \mathrm{AA} \tag{7.5}$$

The rate equations were modeled, resulting in simultaneous differential Eqs. 7.6–7.11.

$$\frac{\mathrm{d[AA]}}{\mathrm{d}t} = -k_{11}[\mathrm{AA}][HP] + k_{12}[\mathrm{PA}][\mathrm{Water}] + k_2[\mathrm{PA}][\mathrm{PS}] \tag{7.6}$$

$$\frac{\mathrm{d[HP]}}{\mathrm{d}t} = -k_{11}[\mathrm{AA}][\mathrm{HP}] + k_{12}[\mathrm{PA}][\mathrm{Water}] \tag{7.7}$$

$$\frac{\mathrm{d[PA]}}{\mathrm{d}t} = +k_{11}[\mathrm{AA}][\mathrm{HP}] - k_{12}[\mathrm{PA}][\mathrm{Water}] - k_2[\mathrm{PA}][\mathrm{PS}] \tag{7.8}$$

$$\frac{\mathrm{d[Water]}}{\mathrm{d}t} = +k_{11}[\mathrm{AA}][\mathrm{HP}] - k_{12}[\mathrm{PA}][\mathrm{Water}] \tag{7.9}$$

$$\frac{\mathrm{d[PS]}}{\mathrm{d}t} = -k_{21}[\mathrm{PA}][\mathrm{PS}] \quad (7.10)$$

$$\frac{\mathrm{d[EPS]}}{\mathrm{d}t} = +k_{2}[\mathrm{PA}][\mathrm{PS}] \quad (7.11)$$

7.3 Results and Discussion

7.3.1 Optimization of the Epoxidation of Palm Stearin

The Taguchi method was employed to determine the optimal conditions for the epoxidation of palm stearin. This approach employs signal-to-noise (S/N) ratio analysis to identify the configurations that yield optimal output with minimal variation. The "larger the better" quality characteristic was employed for the relative conversion to oxirane (RCO), as the objective was to maximize epoxide yield [7, 8, 18]. The average S/N ratios for the three process parameters, reaction temperature (A), stirrer speed (B), and catalyst loading (C) are shown in Fig. 7.1. The analysis of these ratios indicated that the optimal parameters were a reaction temperature of 65 °C (level 1), a stirrer speed of 100 rpm (level 3), and a catalyst loading of 0.4 g (level 3) [19, 20]. The combination of these variables yielded the maximum RCO of 31.79% under the specified experimental conditions. An Analysis of Variance (ANOVA) was performed for clarifying the effects of each process parameter. The results presented in Table 7.4 illustrate the level to which each factor influenced the total variation in the RCO. The ANOVA indicated that catalyst loading was the predominant factor, accounting for an impressive 66.67% of the total contribution [8, 9]. Subsequently, the reaction temperature increased by 32.05%, while the stirrer speed contributed a mere 1.28%, which was negligible..The negligible impact of stirrer speed suggests that mass transfer between the two phases is not the rate-limiting factor under the examined experimental conditions [20, 21]. The significant impact of catalyst loading indicates that the reaction rate is primarily governed by the chemical kinetics of the peracid formation step, catalyzed by sulfuric acid. This signifies that the reaction system is acutely sensitive to the concentration of the acid catalyst, rendering meticulous control over this parameter essential for attaining high yields. The Taguchi and ANOVA analyses collectively determined the optimal process conditions to be a reaction temperature of 65 °C, a stirrer speed of 100 rpm, and a catalyst loading of 0.4 g. A relative conversion to oxirane (RCO) of 31.79% was attained under these conditions [22].

The results from the Analysis of Variance (ANOVA) reveal important insights into the factors affecting the yield of palm stearin epoxidation. Catalyst loading (Factor B) was found to be the most influential parameter, contributing 66.67% to the overall

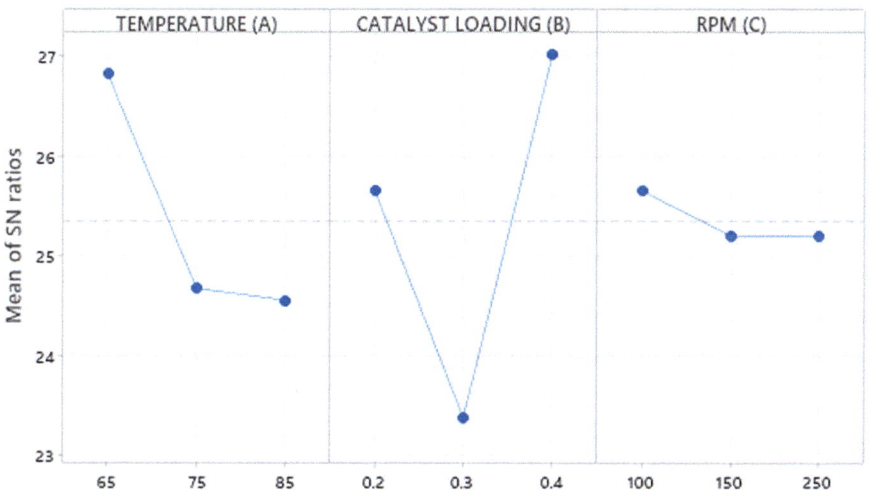

Fig. 7.1 Average S/N ratios for all process parameters affecting the epoxide yield of palm stearin. A: Reaction temperature, B: Stirring speed, C: Catalyst loading, D: Acetic acid/palm stearin molar ratio, E: Hydrogen peroxide/palm stearin molar ratio

Table 7.4 S/N ratio analysis based on the larger the better-quality characteristic

Process parameters and their levels					
No	(A) Reaction temperature (°C)	(B) Catalyst loading (g)	(C) Stirrer speed (rpm)	RCO(%)	S/N ratio
1	65	0.2	100	27.82	28.89
2	65	0.3	150	15.90	24.03
3	65	0.4	250	23.84	27.55
4	75	0.2	150	15.90	24.03
5	75	0.3	250	15.90	24.03
6	75	0.4	150	19.87	25.96
7	85	0.2	250	15.90	24.03
8	85	0.3	100	12.72	22.09
9	85	0.4	150	23.84	27.55

reaction yield. This highlights the critical role of catalyst concentration in accelerating the epoxidation process, as higher catalyst levels provide more active sites for the reaction. However, it is essential to note that after reaching a certain concentration, further increases in catalyst loading result in diminishing returns, suggesting an optimal loading point beyond which additional catalyst has minimal impact [23].

Table 7.5 ANOVA results for epoxidation of palm stearin

Factor	F-Value	P-Value	Percentage contribution (%)
(A) Reaction temperature (°C)	1	0.5	32.05128
(B) Catalyst loading (g)	2.08	0.325	66.66667
(C) Stirrer speed (rpm)	0.04	0.961	1.282051

Reaction temperature (Factor A), which contributed 32.05%, also plays a significant role, as high temperatures typically enhance the reaction rate and promote the formation of epoxide groups. Meanwhile, stirrer speed (Factor C) demonstrated minimal significance, contributing only 1.28%. This suggests that, while stirring ensures homogeneity and effective mixing of reactants, it does not substantially impact the yield once optimal mixing conditions are achieved. The findings from this analysis highlight that for optimizing the epoxidation process, the primary focus should be on adjusting catalyst loading and reaction temperature, with stirrer speed being less critical [24, 25]. This comprehensive understanding provides valuable guidance for refining the reaction parameters to maximize yield and efficiency (Table 7.5).

7.3.2 *In Situ Epoxidation of Palm Stearin*

The mechanism of the in situ epoxidation of palm stearin with peracetic acid is a two-step process. The first step involves the acid-catalyzed reaction between hydrogen peroxide and acetic acid to form the peracetic acid. This reaction is the rate-determining step and occurs predominantly in the aqueous phase. The peracetic acid then acts as an electrophilic species, attacking the carbon–carbon double bonds present in the unsaturated fatty acid chains of the palm stearin triglycerides [26–28]. This electrophilic addition leads to the formation of the strained, three-membered oxirane ring and the regeneration of the acetic acid, which can then re-enter the aqueous phase [18]. The fact that stirrer speed has a minimal effect on the reaction yield, as shown by the ANOVA results, provides a key piece of information about the reaction system. It confirms that under the optimal conditions, the mass transfer of reactants between the aqueous and organic phases is not a limiting factor. The reaction's rate is instead governed by the inherent chemical kinetics of the peracid formation and subsequent epoxidation steps, which are highly dependent on the catalyst and temperature [29]. From Fig. 7.2. The results show a peak in the relative conversion of oxirane (RCO) at 31.79% at 40 min, demonstrating the effectiveness of the reaction at this point under the optimal conditions. The RCO value progressively increased from 7.95% at 10 min to 23.84% at 30 min, marking the initial stages of epoxide formation, with a corresponding rise in the formation of peracetic acid, a crucial intermediate in the epoxidation process [27, 29, 30]. The reaction reached its highest yield at 40 min, with RCO peaking at 31.79%. Afterward, the RCO began to decline gradually, dropping to 19.87% at 50 min, and further decreasing to 7.95% by

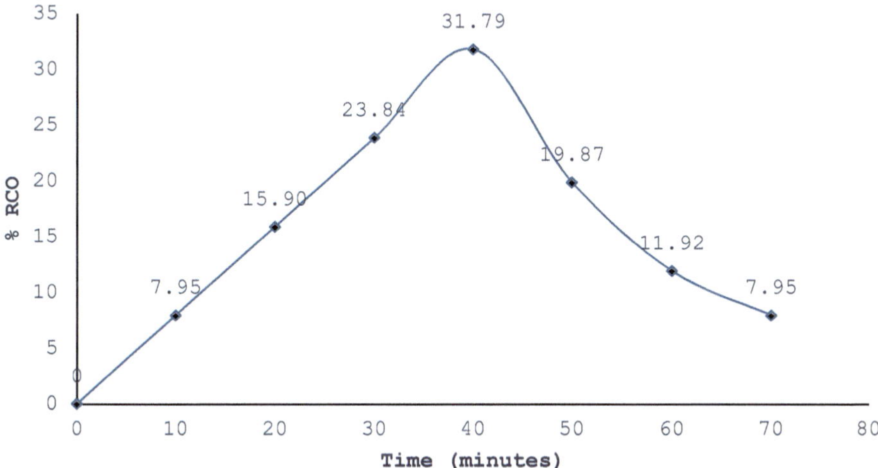

Fig. 7.2 Epoxidation of palm stearin using in situ peracids

the 70-min mark. This reduction suggests a reversible reaction, potentially driven by the breakdown of peracetic acid and the opening of the oxirane ring under prolonged reaction times. These findings are consistent with previous studies for soybean oil epoxidation at approximately 60 °C, where a similar trend of RCO decline was observed after an initial increase [31].

7.3.3 FTIR Analysis of Raw and Epoxidized Palm Stearin

The FTIR spectra of raw palm stearin and epoxidized palm stearin, as shown in Fig. 7.3, show the successful epoxidation. This can be seen from the decrease of the absorption band near 3010 cm^{-1}, indicative of the = C–H stretching of alkene groups; double bonds have been degraded. In the 3000–3050 cm^{-1} range, raw PS exhibits a peak indicative of C–H stretching from unsaturated double bonds, which is absent in epoxide PS, thereby confirming the conversion of double bonds into epoxide groups. Both raw PS and epoxide PS display peaks in the 2920–2850 cm^{-1} range, signifying that the saturated aliphatic chains are unchanged by epoxidation. The 1600–1650 cm^{-1} region shows a notable peak for C = C stretching in raw PS, which is either absent or substantially reduced in epoxide PS, thereby confirming the conversion of double bonds. Furthermore, epoxide PS exhibits peaks at 1240 cm^{-1} and 1097 cm^{-1}, indicative of C–O–C stretching vibrations from the epoxide rings, thereby confirming the presence of epoxide groups [32, 33].

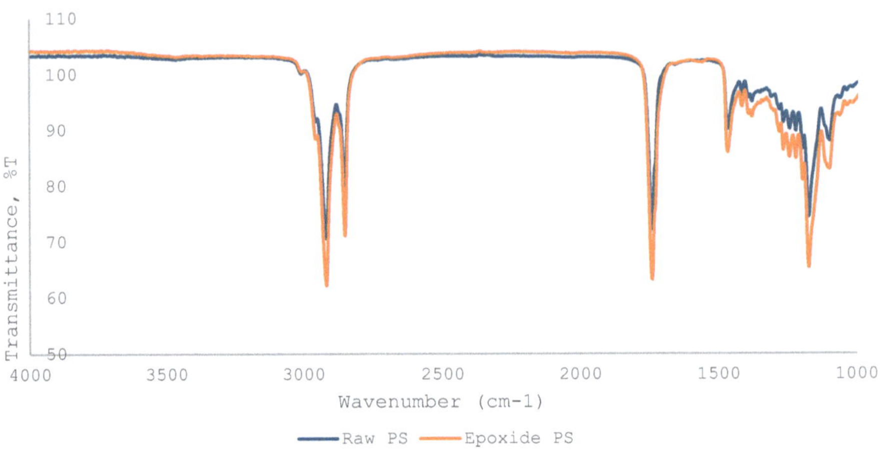

Fig. 7.3 FTIR spectroscopy for raw and epoxidized palm stearin

7.3.4 Kinetic Modeling—Epoxidation of Palm Stearin

To provide a predictive tool for the epoxidation process, a mathematical kinetic model was developed. The model describes the reaction rates of the various species involved in the system. To solve the differential equations that govern the reaction kinetics, the 4th-order Runge–Kutta numerical integration technique was employed [34, 35]. This robust method is well-suited for simulating the non-linear changes in reactant and product concentrations over time, enabling a precise comparison with the experimental data. The kinetic modeling provided specific values for the reaction rate constants, which are crucial for understanding the relative speeds of the reaction steps. These constants are presented in Table 7.6. The values of $k_{11} = 0.0117$ mol/(L·min), $k_{12} = 0.0010$ mol/(L·min), and $k_2 = 0.0106$ mol/(L·min) describe the rates of the individual steps in the proposed reaction mechanism, providing a quantitative basis for the overall process dynamics [36].

The developed kinetic model was validated by comparing its simulated output with the experimental data. The results, illustrated in the provided graph comparing experimental versus simulation data, show a remarkable degree of agreement. This strong correlation is quantitatively supported by an R^2 value of 0.992. This exceptionally high value indicates that the model is a highly accurate representation of the

Table 7.6 Rate constant for epoxidation of palm stearin at optimized process parameters

Reaction rate constant	Value (mol/L·min)
k_{11}	0.0117
k_{12}	0.0010
k_2	0.0106

Correlation coefficient, $r = 0.90$

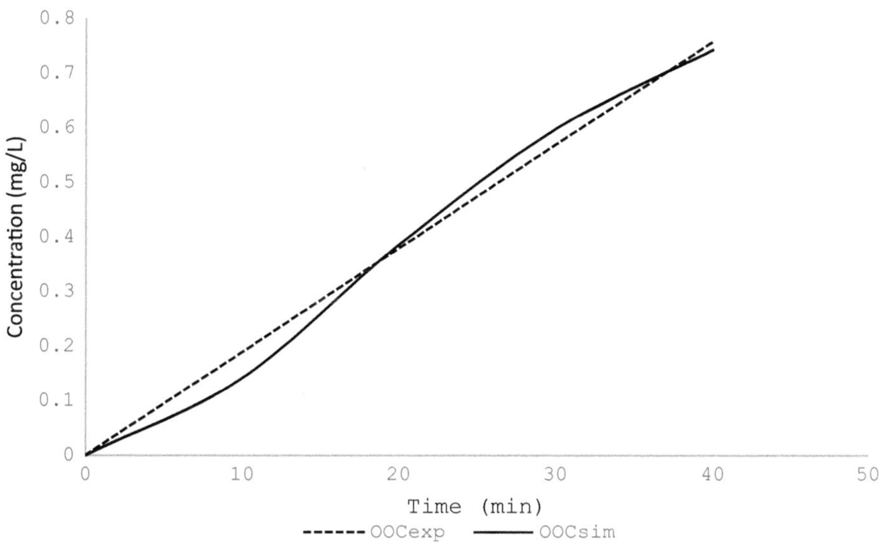

Fig. 7.4 Experimental versus simulation data for epoxidation palm stearin

physical system under the optimal conditions investigated. The high R^2 value is a powerful finding. It suggests that at the determined optimal conditions, the forward epoxidation reaction is highly dominant, and the rates of undesirable side reactions, such as oxirane ring cleavage, are minimal. This contrasts with other studies, such as one on soybean oil epoxidation, which reported a lower correlation coefficient of 0.88 and noted significant deviation at the end of the reaction due to unmodeled degradation reactions [31, 37]. The success of this model implies that the process is highly predictable and controllable under optimized conditions, making it an ideal candidate for industrial-scale simulation and implementation. The model can be used confidently to predict the yield and reaction progress, significantly reducing the need for extensive experimental trials (Fig. 7.4).

7.4 Conclusion

This study successfully demonstrates the optimization and kinetic modeling of the sulfuric acid-catalyzed in situ epoxidation of palm stearin. The Taguchi experimental design shows the optimal process parameters: a reaction temperature of 65 °C, a stirrer speed of 100 rpm, and a catalyst loading of 0.4 g. A maximum relative conversion to oxirane (RCO) of 31.79% was attained under these conditions. From the Analysis of Variance (ANOVA) indicated that catalyst loading was the predominant factor, significantly contributing 66.67% to the epoxide yield. The stirrer speed had a negligible impact, accounting for merely 1.28%. The results indicate that the reaction rate is predominantly determined by the inherent chemical kinetics of peracid formation, rather than by mass transfer constraints between the aqueous and organic phases. A predictive mathematical model was designed utilizing the 4th-order Runge–Kutta method, which exhibited a robust correlation with the experimental data. The model's elevated coefficient of determination ($R^2 = 0.992$) validates its accuracy and indicates that, under optimized conditions, the epoxidation reaction predominates, rendering side reactions like oxirane ring cleavage insignificant. This study's results offer a thorough framework for the efficient and predictable production of epoxidized palm stearin, a significant bio-based raw material. The established kinetic model functions as a reliable instrument for process regulation and offers a fundamental framework for prospective industrial-scale implementations of this sustainable synthesis technique. This study enhances oleochemical processes, providing an eco-friendly substitute for petroleum-derived feedstocks and converting an industrial by-product into a valuable product.

References

1. B. Abdullah, R. Yusop, J. Salimon, D. Derawi, W. Ahmed, Epoxidation synthesis of linoleic acid for renewable energy applications. Malaysian J. Analyt. Sci. **20**(1), 131–141 (2016). https://doi.org/10.17576/MJAS-2016-2001-14
2. F.K.A. Bustamam, C.B. Yeoh, N. Sulaiman, M.H. Saw, Evaluation on the quality of Malaysian refined palm stearin. Oilseeds and Fats, Crops and Lipids **29**, 37 (2022). https://doi.org/10.1051/ocl/2022030
3. S. Kanagaratnam, N.L.H.M. Dian, R.A. Hamid, N.H. Ismail, W.R.A. Isa, N.A.M. Hassim, Z. Omar, M.M. Sahri, Are characteristics of soft palm stearins similar to soft palm mid fractions? J. Oil Palm Res. (2020). https://doi.org/10.21894/JOPR.2020.0004
4. S. Dworakowska, D. Bogdal, A. Prociak, Microwave-assisted synthesis of polyols from rapeseed oil and properties of flexible polyurethane foams. Polymers (Basel) **4**(3), 1462–1477 (2012). https://doi.org/10.3390/polym4031462
5. Z. Zhu, J.H. Baker, C. Liu, M. Zhao, M. Kotaki, H. Sue, High performance epoxy nanocomposites based on dual epoxide modified α-Zirconium phosphate nanoplatelets. Polymer **12**, 123154 (2020). https://doi.org/10.1016/j.polymer.2020.123154
6. E. Purwanto, *The Synthesis of Polyol from Rice Bran Oil (RBO) Through Epoxidation and Hydroxidation Reactions* (Master's thesis, University of Adelaide) (2010)
7. A.A. Sipaut, C.S. Sundang, M. Saalah, T.C. Hoon, M.N.M. Ibrahim, I.A. Rahman, Abdullah, Synthesis and characterization of polyol from refined cooking oil for polyurethane foam formation. Cell. Polym. **31**(1), 19–38 (2012)
8. M.B. Mahadi, I.S. Azmi, M.A.A.H.M. Tajudin, E.A. Saputro, M.J. Jalil, Sustainable epoxidation of sunflower oil via heterogeneous catalytic in situ peracids mechanism. Biomass Conversion and Biorefinery **15**(4), 5719–5727 (2025)
9. N.H. Mudri, L.C. Abdullah, M.M. Aung, M. Rayung, Comparative study of aromatic and cycloaliphatic isocyanate effects on physico-chemical properties of bio-based polyurethane acrylate coatings. Polymers (Basel) **12**(7), 1–17 (2020)
10. I.M. Rasib, N.M. Mubarak, I.S. Azmi, M.J. Jalil, Epoxidation of neem oil via in situ peracids mechanism with applied ion exchange resin catalyst. Biomass Conversion and Biorefinery **15**(9), 13707–13717 (2025)
11. S. Ramanujam, C. Zequine, S. Bhoyate, B. Neria, P. Kahol, R. Gupta, Novel biobased polyol using corn oil for highly flame-retardant polyurethane foams. C, **5**(1), 13 (2019). https://doi.org/10.3390/c5010013

12. A. Banan, H. Mehdipour, Controlled degradation and functionalization of natural rubber by ozonolysis in organic solvent. J. Polym. Res. **28**(9), 1–13 (2021). https://doi.org/10.1007/s10965-021-02671-2
13. P. Alagi et al., Functional soybean oil-based polyols as sustainable feedstocks for polyurethane coatings. Ind. Crops Prod. **113**, 249–258 (2018). https://doi.org/10.1016/j.indcrop.2018.01.041
14. C. Chang, R. Ding, M.R. Kessler, Reduction of epoxidized vegetable oils: a novel method to prepare bio-based polyols for polyurethanes. Macromol. Rapid Commun. **35**(11), 1068–1074 (2014). https://doi.org/10.1002/marc.201400161
15. I. Boustead, *Eco-profiles of the European Plastics Industry—Polyols*. Eco-profiles of the European Plastics Industry-POLYOLS, pp. 1–19 (2005).
16. M.Z.A. Kadir, I.S. Azmi, N.D. Kasmin, S.J.A. Rahman, M.J. Jalil, Green catalytic epoxidation of hybrid oleic acid derived from waste palm cooking oil + palm oil. Polym. Bull. **81**(8), 6979–6994 (2024)
17. T.A.Z. Tunku Ozir, M.Z.B. Ab Kadir, I.S. Azmi, M.Z. Yeop, S.M.A. Rahman, M.J. Jalil, Bio-lubricant production based on epoxidized oleic acid derived from dated palm oil using in situ peracid mechanism. Int. J. Chem. Reactor Eng. **21**(6), 793–800 (2022). https://doi.org/10.1515/ijcre-2022-0161
18. M. Musik, M. Bartkowiak, *Advanced Methods for Hydroxylation of Vegetable Oils, Unsaturated Fatty Acids and Their Alkyl Esters* (2022)
19. T. Vlček, Z.S. Petrović, Optimization of the chemoenzymatic epoxidation of soybean oil. J. Am. Oil. Chem. Soc. **83**, 247–252 (2006). https://doi.org/10.1007/s11746-006-1200-4
20. S.J.A. Rahman, M.A. Rahman, N. Hambali, I.S. Azmi, M.J. Jalil, Sustainable approach for catalytic epoxidation of oleic acid followed by in situ ring-opening hydrolysis with applied ion exchange resin. Int. J. Chem. Reactor Eng. (2024). https://doi.org/10.1515/ijcre-2023-0196
21. A. Fridrihsone, F. Romagnoli, V. Kirsanovs, U. Cabulis, Life cycle assessment of vegetable oil-based polyols for polyurethane production. J. Clean. Prod. **266**, 121403 (2020). https://doi.org/10.1016/j.jclepro.2020.121403
22. J.K.B. Balangao, A.A. Lubguban, R.J.G. Ruda, R.H. Aquiatan, S. Paclijan, Soy-based polyols and polyurethanes. Kimika, **28**(1), 1–19 (2017). https://doi.org/10.26534/kimika.v28i1.1-19
23. F. Marriam, A. Irshad, I. Umer, M. Arslan, Vegetable oils as bio-based precursors for epoxies. Sustain. Chem. Pharm. **31**, 100935 (2023). https://doi.org/10.1016/j.scp.2022.100935
24. B. Nascimento, A.P. Costa, Hydroxylation of sunflower, corn and crambe oils and chemical characterization of the obtained vegetable polyols. J. Mod. Chem. Chem. Technol. **25**, 23–28 (2020)
25. E. Aydoğmuş, F. Kamişli, New commercial polyurethane synthesized with biopolyol obtained from canola oil: Optimization, characterization, and thermophysical properties. J. Mol. Struct. **125610**, 132495 (2022). https://doi.org/10.1016/j.molstruc.2022.132495
26. S. Silviana, D. Anggoro, A. Kumoro, Kinetics study of waste cooking oil epoxidation with peroxyacetic acid using acid catalysts. Rasayan J. Chem. **12**(03), 1369–1374 (2019). https://doi.org/10.31788/rjc.2019.1235190
27. S. Sinadinović-Fišer, M. Janković, Z. Petrović, Kinetics of in situ epoxidation of soybean oil in bulk catalyzed by ion exchange resin. J. Am. Oil. Chem. Soc. **78**(7), 725–731 (2001). https://doi.org/10.1007/s11746-001-0333-9
28. H.H. Habri, I.S. Azmi, N.M. Mubarak, M.J. Jalil, Auto-catalytic epoxidation of oleic acid derived from palm oil via in situ performed acid mechanism. Catalysis Surveys from Asia (2023)
29. J. Peyrton, L. Avérous, Oxazolidone formation: Myth or fact? The case of biobased polyurethane foams from different epoxidized triglycerides. Polym. Chem. **12**(1), 143–155 (2021). https://doi.org/10.1039/D1PY00129A
30. X. Cai, J. Zheng, A. Aguilera, L. Vernières-Hassimi, P. Tolvanen, T. Salmi, S. Leveneur et al., Influence of ring-opening reactions on the kinetics of cottonseed oil epoxidation. Int. J. Chem. Kinet. **50** (10), 726–741 (2018). https://doi.org/10.1002/kin.21208
31. A. Shahrizan, N. Hanib, I. Azmi, M. Fauziyah, M. Jalil, In situ epoxidation of canola oil via peracetic acid mechanism—optimization and kinetic study. J. Elastomers Plast. **56**(4), 454–468 (2024). https://doi.org/10.1177/00952443241243376

32. A. Osemeahon, S. Okoye, Synthesis, physicochemical, and FTIR characterization of vegetable oil-based polyol from Ximenia americana seed oil. Ch&ChT **9**(3), 23–28 (2018)
33. N. Thủy, Kinetic study of epoxidation of rubber seed oil using tungstate-based catalyst. VNU J. Sci. Nat. Sci. Technol. **34**(4) (2018). https://doi.org/10.25073/2588-1140/vnunst.4799
34. G. Ruihua, C. Ma, S. Sun, Y. Ma, Kinetic study on oxirane cleavage of epoxidized palm oil. J. Am. Oil. Chem. Soc. **88**(4), 517–521 (2010). https://doi.org/10.1007/s11746-010-1697-4
35. P.D. Jadhav, A.V. Patwardhan, R.D. Kulkarni, Kinetic study of in situ epoxidation of mustard oil. Mol. Catalys. **511**, 111748 (2021). https://doi.org/10.1016/j.mcat.2021.111748
36. M. Jalil, N.H.A. Rasnan, A.F.M. Yamin, M.S.M. Zaini, N. Morad, I.S. Azmi, M.B. Mahadi, M.Z. Yeop, Optimization of epoxidation palm-based oleic acid to produce polyols. Chem. Chem. Technol. **16**(1), 66–78 (2022). https://doi.org/10.23939/chcht16.01.066
37. Z. Petrović, A. Zlatanić, C. Lava, S. Sinadinović-Fišer, Epoxidation of soybean oil in toluene with peroxoacetic and peroxoformic acids—kinetics and side reactions. Eur. J. Lipid Sci. Technol. **104**(5), 293–299 (2002)

MIX
Papier aus verantwortungsvollen Quellen
Paper from responsible sources
FSC® C105338

If you have any concerns about our products,
you can contact us on
ProductSafety@springernature.com

In case Publisher is established outside the EU,
the EU authorized representative is:
Springer Nature Customer Service Center GmbH
Europaplatz 3, 69115 Heidelberg, Germany

Printed by Libri Plureos GmbH
in Hamburg, Germany